D0955579

FIRE *and* FLOOD

FIRE

and

FLOOD

*A People's History of Climate Change,
from 1979 to the Present*

EUGENE LINDEN

PENGUIN PRESS

NEW YORK

2022

PENGUIN PRESS
An imprint of Penguin Random House LLC
penguinrandomhouse.com

LIBRARY OF CONGRESS CATALOGING-IN-PUBLICATION DATA
Names: Linden, Eugene, author.
Title: Fire and flood: a people's history of climate change, from
1979 to the present / Eugene Linden.
Description: New York: Penguin Press, 2022. |
Includes bibliographical references and index.
Identifiers: LCCN 2021021241 (print) | LCCN 2021021242 (ebook) |
ISBN 9781984882240 (hardcover) | ISBN 9780593295724 (ebook)
Subjects: LCSH: Climatic changes. | Climatic changes—Government policy. |
Climatic changes—Economic aspects.
Classification: LCC QC903 .L585 2022 (print) |
LCC QC903 (ebook) | DDC 304.2/8—dc23
LC record available at https://lccn.loc.gov/2021021241
LC ebook record available at https://lccn.loc.gov/2021021242

Printed in the United States of America
1   3   5   7   9   10   8   6   4   2

Designed by Amanda Dewey

To the memory of those prophetic scientists who early on warned of the enormous costs of tampering with the atmosphere, a group that includes Wallace Broecker, Sherwood Rowland, Ralph Cicerone, Paul Crutzen, Stephen Schneider, and Hans Oeschger, among many others; and also to those scientists still living who were first responders to the climate crisis, including George Woodwell, Mario Molina, and Veerabhadran Ramanathan; and finally to those scientists, notably Michael Mann, who fought back against the smears and harassment inflicted by climate deniers.

We should have listened to all of you.

What is the use of having developed a science well enough to make predictions if, in the end, all we're willing to do is stand around and wait for them to come true?

—F. SHERWOOD ROWLAND in his speech accepting the Nobel Prize in Chemistry in 1995

CONTENTS

PREFACE

C all it what you like, climate change or global warming, it's here and it's going to get worse, very likely far worse. Sixty years from now, people may wonder what happened in the past that allowed this nemesis to ruin the world. What were the decisions, or lack thereof, the missed opportunities, the political failures that caused a technologically advanced civilization to continue to alter earth's climate even as its leaders knew better? These are the questions this book will attempt to answer.

More than four billion people have been born since climate change became a mainstream issue in 1988. More than five billion people have been born since the issue first appeared on President Jimmy Carter's agenda. Billions of people, now well into adulthood, have lived their entire lives under the specter of global warming. During the past decade, average global temperatures crept up more than 1 degree Celsius over average levels before the Industrial Revolution. In fact, average temperatures are now regularly over the upper bound of what has prevailed since the dawn of the Holocene epoch—the period, starting 11,600 years ago, during which a human-friendly climate nurtured the dawn of civilization and allowed human numbers to grow from roughly five million to nearly eight billion. Noting this inflection point, renowned climate modeler Stefan Rahmstorf put up a sardonic post on Facebook congratulating humans because "we have now left the Holocene." Several hundred million people have recently been born into a new climate epoch.

Roughly one-third of the world's population is unaware that global warming is happening, and the majority of those who are aware that we are changing the climate are either insufficiently alarmed at the prospect or already too fatalistic.

So here we are, in the first innings of a rapid climate change event that is already getting very costly. Was this predicted? Yes, perhaps more than any disaster in history. Could this have been averted? This turns out to be a charged question. We have had more than thirty years of warnings, innumerable conferences and scientific gatherings, even ratified treaties on preventing a climate catastrophe. In recent years, however, the idea that we never had a chance has gathered strength. This argument holds that in the early 1990s, when the scientific consensus on the threat solidified, renewables were too expensive and inefficient to replace fossil fuels and, furthermore, that people did not consider global warming to be an imminent threat, which meant that there was no public alarm to propel political action (a mobilized populace being a necessary precondition to break the hold of fossil fuels on the global economy).

These are not trivial arguments, but they aren't true, as I will explain. We could have taken action, and it's important to understand why we didn't, because the reason points directly to what is likely to happen next and what we can do about it.

Many factors, of course, explain society's response to the discovery that our own actions are changing the climate. Looming over all of them has been the role of business and finance. Indeed, the business world has been the master puppeteer in terms of influencing public awareness, politicians, even the presentation of the science. Looking back at the history of the modern climate change era, we see that if the business community does not want something to happen, it usually doesn't.

Obviously, there are many competing interests in the business community. That said, the record shows which voices were dominant at critical points during the climate change era. During most of that time, the dominant messaging came from the wide spectrum of interests related to fossil fuels with the support of the even broader community of business interests who shared an antiregulatory bias. As we will see, these voices outweighed the discoveries of scientists and the ever more alarming signals coming from the climate itself. From the beginning, business interests have proved adept at reframing the issue, dismissing the risks, demonizing the scientists, and defaming those seeking action as elitist dilettantes who want to tell you what to do and take away your job.

The posture of the business community also figured in the biggest missed opportunity of the climate change era, as the giant emerging economies were making decisions at the beginning of their push for industrial development and expansion of their energy infrastructures. Given a choice between renewables and fossil fuels, almost all chose coal. The greenhouse gas (GHG) emissions of the massive buildout that followed have much more than outweighed all the efforts (such as they've been) of the developed nations to reduce their dependence on carbon to power their economies. Indeed, in 2019,

China alone emitted more greenhouse gases than all thirty-eight developed nations in the Organization for Economic Cooperation and Development (OECD) combined.

The one thing the business and finance community could not and cannot influence is reality itself, although it has proved adept at distorting the public perception of the impacts of changing climate that have been staring us in the face. As these impacts have become more frequent and intense, they have broken through the fog of disinformation disseminated by fossil fuel interests. In fact, the reality of climate change has reached a point where it has altered the posture of much of the business community toward the threat. The question of the day is whether those alterations will be sufficiently broad that this immensely powerful sector can become part of the solution and help reverse the momentum of climate change before it is too late.

For those who don't think climate change is a problem, this book is not for you. I'm not going to try to convince you the threat is real (and if you aren't convinced by now, I'm not sure anything will convince you), but I'll offer a suggestion as to how you might spend your time instead of reading this book: start an insurance company and sell flood and storm insurance policies in all those coastal communities where traditional insurers are pulling out. If you're right about climate change, you'll make a killing.

I come to this project having followed the climate change story since the 1970s and having written about it since the late 1980s. I've written essays, articles, and one prior book about climate change. Some of the essays and many of the articles were for *Time* magazine in the 1980s and 1990s. I've tried to get the word to the broader public through op-eds and articles in massive circulation vehicles such as *Parade* magazine (back in 2006 when people read the news on paper and *Parade* had a massive circulation); I've participated in a number of documentaries and dozens of conferences; and I've given talks at scores of institutions. I've covered global warming from a wide variety of approaches, exploring its impact on the spread of disease and its potential financial consequences. In my reporting on the issue, I've traveled to the Arctic and Antarctica and many places in between. All of this is to say that climate change has been my companion for more than three decades. I've witnessed firsthand how the climate change story unfolded since it became a mainstream issue more than thirty years ago.

I will draw on all this history in an attempt to understand how we got to this point where a problem that seemed so far in the future when I first wrote about it has become a clear and present danger for all humanity, not to mention countless other life forms with whom we share our fragile planet. And I will draw on other experiences and investigations into the nature of markets and the consumer society to show a narrow path that might yet get us out of the mess we have created.

INTRODUCTION

There have been several false dawns in our efforts to tackle the threat of global warming. President Jimmy Carter convened a blue-ribbon panel to study the problem of carbon dioxide emissions in the late 1970s. The group, which included pioneering climate scientists Roger Revelle and George Woodwell, presented Carter's Council on Environmental Quality with a paper in 1979, entitled *The Carbon Dioxide Problem*, which warned that if we didn't take action to curb greenhouse gas emissions, we would see changes in climate by the end of the twentieth century.

Another false dawn broke nine years later during the sweltering summer of 1988, prompted by James Hansen's dramatic testimony to a U.S. Senate committee in which he argued that global warming had already begun. Out of that wave of alarm came a toothless agreement called the Kyoto Protocol of 1997, which went into effect in 2005. Most countries ratified it (although not the United States), but it did little to halt the rise in emissions of greenhouse gases. There have been several pulses of public interest since, the most recent false dawn being the Paris Agreement on climate change of 2015, which was signed by every nation on the planet (until the United States pulled out), but none of them has stopped greenhouse gas emissions from inexorably rising, now more than 60 percent above the levels of 1990, when the global community first began to talk about taking steps to halt their growth.

Now, a real dawn of climate action finally seems at hand. Just six years after the Paris Agreement became a treaty, things are changing fast. Not as fast, alas, as the climate itself. Many of the climate-related changes we're living through—rising seas, more intense heat waves, floods, droughts, massive brush and forest fires around the world, plagues of locusts and outbreaks of pathogens, record warm years, melting permafrost, and changes in the ice sheets—first became noticeable in the 1990s. What's different today is that they are coming so thick and fast, and are producing such outsized costs, that they cannot be ignored. Before the 1990s, Australia, for instance, experienced severe drought roughly every eighteen years. Since then, both the severity and length have been increasing. Australia was plagued by recurring droughts from 2001 until 2009. Drought, heat, and wind caused devastating bushfires in 2009 that destroyed two thousand homes, displaced seventy-five hundred people, and killed an estimated one million animals. Drought resumed in 2017 and has continued since, culminating in 2019 with some of the worst wildfires in the country's history, during the country's hottest and driest year on record. This time around, the fires killed an estimated 1.25 billion animals, more a than a thousand times the estimated wildlife casualties of the 2009 Black Saturday fires. More than a billion more animals were killed as the fires continued in 2021.

Extreme weather events are coming with increasing frequency and greater intensity. In 2017, Hurricane Irma knocked out power for one million people in Puerto Rico; two weeks later, Hurricane Maria hit, cutting power for the rest of the island. The storms killed nearly three thousand people, and the island has yet to recover. Five of the ten most damaging wildfires in California history took place in 2020 alone. Then, in 2021, California had its largest fire in the state's history. For a period in July and August of that year, it seemed like the whole world was on fire, with scores of wildfires burning in Greece and Turkey, and some of the largest fires ever seen incinerating millions of acres of Taiga forest in the Russian Far East. Smoke from these fires cast a pall that extended thousands of miles from the fires themselves.

None of this is news to anyone with a passing interest in current events. While the pace of this tide of extreme weather events has accelerated, the flood of data points has been flowing for three decades. During most of that period, the connection between extreme weather and climate change didn't seem to matter to the public or the business community. Now it does.

That's because things are getting expensive. According to a study undertaken by the giant insurance broker Aon, weather-related disasters around the world led to $1.8 trillion in losses between 2000 and 2010. The figure for 2010 to 2019 is $3 trillion. Monsoon floods in China wreaked $15 billion in damage in 2019; on the other side of the globe, the Mississippi floods inflicted $10 billion in losses.

In the United States, the belated awakening to the threat is just the latest iteration of a deeply ingrained response to threats that has become typical in America. Warnings, whether they be of a financial crash, a pandemic, a threat of pollution or climate change, go unheeded until a crisis hits, and then politicians and the public react. The cycle is so predictable that social and economic commentators even have a phrase for it, "barn door closing," meaning that we tend to take action only after the cows have fled. Thus, only after the financial crisis of 2008 did Congress enact the Volcker Rule to prevent banks from taking the risks that almost brought down the global financial system (and then as memories faded in the ensuing years, Congress quietly rolled back those protections, setting the stage for the next crisis). This pattern of barn door closing is not an accident. The reason we usually don't act until a crisis has occurred is that business as usual has enormously powerful momentum in our society, and until a crisis happens, the danger can be dismissed as speculative.

A threat to the atmosphere discovered in the 1970s provides a case study of how the business community reacted to a newly discovered hazard, foreshadowing the response of moneyed interests to the later discovery of global warming.

This threat was the discovery that certain chemicals were damaging the ozone layer. The global response has been cited as a great environmental success story, but the truth is more mixed. Sherwood Rowland and Mario Molina first warned of the danger that the unchecked release of chlorofluorocarbons (CFCs) posed to the ozone layer in 1974 (in 1995, along with Paul Crutzen, they were awarded the Nobel Prize in Chemistry for their groundbreaking work), but the global community did not act for another fifteen years. Even after it was discovered that CFCs were responsible for knocking a giant hole in the ozone layer, it took another four years for the global community to take action.

The Montreal treaty was a great environmental victory only if we ignore one small issue: once CFCs get into the upper atmosphere, they stay there for decades. Thus, if there is no backsliding on the moratorium on producing CFCs—not at all a certainty—the ozone layer will not heal until the 2030s, roughly six decades after the problem was first identified.

Climate change poses a far more difficult challenge than CFCs. The threat to the ozone layer came from one class of chemicals used in specialized activities. With climate change it sometimes seems as though almost all economic activity contributes to the problem—agriculture, transportation, power generation, concrete and steel production, construction, you name it. And greenhouse gases have an even longer life in the atmosphere than some CFCs.

Moreover, at least at first, the impacts of climate change seemed to unfold so intermittently and so incrementally that each anomalously strong storm, drought, flood, or heat wave could be written off as just that: an anomaly and not a warning that the world was changing. Sure, the United States sweltered in the summer of 1988, but then Mount Pinatubo erupted in June 1991, throwing up a globe-circling volcanic haze, and the following year the Northeast had the second-coldest summer on record, with snow in June in New York. In later chapters, we'll look more closely at how the episodic and incremental aspects of a changing climate make it easy to ignore various climate impacts right up until the point they become catastrophic. As Sherwood Rowland remarked when he accepted the Nobel Prize, "What is the use of having developed a science well enough to make predictions if, in the end, all we're willing to do is stand around and wait for them to come true?"

Warnings go unheeded because it is deeply ingrained in capitalism to disregard any action that might interrupt the existing flow of money, until a crisis can no longer be ignored. This inherent vice of capitalism, especially the American way of business, sets up a tragic collision of geophysics and economics: climate change demands proactive measures from a system that is geared to be reactive. The result is that the world has loaded more greenhouse gases into the atmosphere in the thirty-four years since climate change became an international issue than it did in the entire industrial age up to that point. As noted, a good portion of that increase has come from China, which in 2017 put ten times more carbon into the atmosphere than it did in 1972.

Despite the mismatch between the timetables of capitalism and the timetables of climate change, there were and are levers in the economy that could have permitted proactive measures in the 1990s to reduce our dependency on fossil fuels. One of the great trends of the past century has been the shift in almost all countries from rural to urban. Given the proper infrastructure, a major city might transition from coal to hydro or renewables with the flick of a switch. And while greenhouse gas emitters number in the millions, just a few score companies account for the majority of fossil fuel production and use.

The most reasonable argument for why little action was taken in earlier decades is that renewable technologies were too expensive to provide a realistic alternative for power generation and transportation. But to say that also begs a question: Why were alternative energy sources so underdeveloped in the 1990s?

Viewed in a historical context, the question is damning. Inventors have been looking at how to capture solar, tidal, and other forms of natural energy since our ancestors first began to speculate on ways to enhance human power. Humans are no more intelligent now than they were 10,000 years ago (or 150,000 years ago for that matter), and long before the discovery of cheap oil in the Arabian desert, inventors saw the power of the sun, the wind, the tides, and the heat trapped in the earth and have tried to imagine ways to capture or divert that energy to human use. And long before James Hansen spoke to Congress about global warming in 1988, some thinkers wondered whether we might be forced to turn to alternative sources of energy because of fossil fuels' potential impact on the climate.

Writing for the Smithsonian Institution, Robert Thurston argued that even before fossil fuels are exhausted, emissions from smokestacks may alter climate, and that ultimately humans may be forced to harness the energy of the sun. That article was published in 1901. Back then it would have been natural to expect that renewables would be the major source of power a hundred years in the future. Indeed, in 1900, the *Boston Globe*, in an article on what life would be like in the year 2000, offered the following prediction of a time "when the tides in the harbor will be made to furnish heat, light and power, when every person will own his own automobile, or whatever it may be called in that day." They were right about the automobile. They could not have been more wrong about what would be powering Boston.

Thurston wrote at the end of the Victorian era, a period of great invention and technological hubris, but before oil gained its stranglehold on the economy. It was a period boiling with ideas about powering industry. Many of the innovations being investigated today amount to dusted-off ideas of a century earlier. Or even earlier still.

For instance, some solar projects in the American Southwest convert heat generated by concentrated solar energy into electricity in part through a Stirling engine, a device invented by the Reverend Robert Stirling in 1816. A 400-megawatt power plant in the Mojave Desert uses the same basic principles as a solar power system built in Egypt in 1913 by an engineer named Frank Shuman to drive pumps on an irrigation system. Both rely on an array of mirrors to concentrate the sunlight into usable heat.

Water is 850 times as dense as air, and it has occurred to many inventors that the immense power of moving water, whether it is moving as a current, a tide, or in waves, might power turbines to generate electricity. In the late nineteenth century, Thomas Edison came up with the idea of mounting impellers on the ocean floor off the coast of Florida to capture the tremendous energy of the Gulf Stream, which moves 8 billion gallons a minute as it approaches within 15 miles of the East Coast. Today, Verdant Energy applies the same concept to produce electricity from the fast-moving, tide-driven East River in New York City, and projects producing energy from rivers and tides are generating power in bodies of water around the world.

Wind has been harnessed by sailors and millers for thousands of years, and wind-powered electrical generation dates back to the nineteenth century. In 1938, some of the scientists who worked on the first atomic bomb built a 1-megawatt wind turbine in Vermont. Human capture of geothermal power for heat dates back ten thousand years, and its use for electrical generation dates to 1904 when Prince Piero Ginori Conti of Trevignano built a 10-kilowatt plant in Larderello, Italy. (Descendants of that plant now generate 800 megawatts of electricity.)

What derailed renewables in 1900? Something better came along. That was oil. In the decades following 1859, when oil was first found in Pennsylvania, a series of vast new discoveries made the fuel ever cheaper, and improvements in refining made it ever more useful. It turned out that oil or its by-products could do almost anything one could do with coal, and more. Compact, packed with energy, and easy to move around, oil and gas derivatives came to dominate transportation fuels and utterly perfused the economy. The more the world came to depend on it, the more its strategic importance pushed alternatives to the sidelines. Starting around 1908, when vast reserves were discovered in what is now Iran, and accelerating in 1938, when what seemed to be unlimited supplies of oil were discovered in Saudi Arabia, renewables fell into a coma, where they remained for decades.

The U.S. space program gave something of a boost to solar power research, and then in 1973, renewables caught a big break, thanks to the hubris of the OPEC nations, which, in response to U.S. support of Israel during the Yom Kippur War, imposed an embargo on exports to America and other countries. The embargo produced huge lines of cars waiting at gas pumps, sent oil prices skyrocketing in the United States, and exposed our dependence on Middle East suppliers. But that effort fizzled when Ronald Reagan came into office.

Consequently, solar, tidal, and, to some degree, wind power were behind the curve when the first real false dawn arrived, the clamor to limit greenhouse gases that arose in the years before and after 1990. At that time, it was fossil fuels, not renewables, that received the lion's share of government subsidies (and still do, to the tune of $400 billion annually around the globe). Those who argued in 1990 that there were no viable alternatives to fossil fuels except hydropower and nuclear power had half a point, but it was analogous to a football coach who shoots a sub in the leg and then benches him because he's not ready to play.

In the years since 1990, investments in solar, wind, tidal, geothermal, and other renewables have ramped up. Today, wind, solar, geothermal, and even tide power are cost competitive with fossil fuels in many markets around the world. Wind power now accounts for roughly 20 percent of the electricity generated in Texas and is so plentiful and cheap that the giant utility TXU has a program that gives electricity away for free after nine p.m. to some of its customers. This has happened despite the best efforts of the George W. Bush and Donald Trump administrations to support fossil fuels, particularly coal, and to derail efforts to promote renewables.

When economists argue that fossil fuels are essential to our prosperity, they are also saying, implicitly, that luck, not ingenuity, fueled the growth of the Industrial Revolution. We'd better hope that's not true, and there is every reason to believe that it isn't. The same ingenuity that found multiple ways to expand and optimize the utility of fossil fuel combustion since the nineteenth century might have gone into improving wind turbines and tidal impellers. Material sciences might have focused on the best materials for converting sunlight much earlier than they did.

Industry might have found less carbon-intensive ways to produce basic materials. A perfect case in point comes from the steel industry, where the smelting of 2 billion tons of global production accounts for 8 percent of CO_2 emissions. As documented by Maria Gallucci of *Grist*, MIT scientists invented a way to make steel with zero GHG emissions by passing an electrical current through iron oxide and other metals, and the heat ultimately produces steel with an exhaust residue of oxygen. The scientists formed a company called Boston Metal to produce steel, and they argue that they can do so at costs competitive with coal-fired plants.

The process requires no new high-tech breakthroughs. Indeed, *Grist* quotes Donald Sadoway, the MIT material chemistry professor, as saying that he got the idea decades ago as an alternative way of making aluminum. The basic materials of this process—non-fossil-fuel electricity, iron oxide, and other metals—have been available for more than a century, and the only reason this process was not invented sooner might be described as lack of motivation. With coal as a dirt-cheap energy source, steel makers had little incentive to look for alternative ways to make it.

Had not cheap oil driven renewables to the fringes for several decades there is no question that solar, wind, geothermal, and tidal would have become economically viable much sooner. That this did not happen was not destiny, just another example of the self-perpetuating momentum of a technology once it has become the standard. No technological paradigm has perfused the modern economy more thoroughly than those related to the combustion of fossil fuels (though information technology will likely become more pervasive in the coming decades).

Unfortunately, climate change has its own timetable governed by geophysics, not business and politics, and it has ramped up even faster. Greenhouse gases have their own momentum. Once carbon dioxide is created through combustion, it remains in the atmosphere for many decades. Even if the world cut emissions to zero tomorrow, which it can't, further climate change is locked in, and the problem is likely to become ever more disruptive for decades to come. Increased volatility will arise from the second-order impacts set in motion by the warming we've had so far. These include the melting of the permafrost, which is releasing increasing amounts of CO_2 as well as methane, an extraordinarily potent greenhouse gas, and the immensely important but still uncertain response of the oceans to the heat they have absorbed thus far.

The story of how we got here requires the unpacking of four different realms and how they have interacted over the past three decades. The first realm is reality itself: what has actually happened. The second is the state of science. The third is public awareness. The fourth is the world of finance and industry. I don't include politics as a fifth realm because politicians have largely been reactive, and their actions have been derivative, driven by the mood of the public and, to a much greater degree, by the realm of money, where the real power lies.

The actions of this fourth realm, the world of business, finance, and industry, explain why the United States and the rest of the world have been so unprepared for a warming globe. For most of the climate change era (which I date from 1988, when it became a mainstream issue), the aggregate impact of the financial/business community was to delay action on the problem, even if many investors and executives acknowledged that the danger was real. The implicit cost-benefit analysis was that regulation and other preventive measures posed an immediate threat to profits, while the costs of climate change lay far in the future. Acting on that analysis, the business/ investing community devoted enormous resources to sowing confusion about the issue in the public and throwing sand in the gears of any nascent steps toward action.

This is not news. The scope and details of these efforts have been exhaustively documented in articles, books, and documentaries. The business/financial community, however, is not a monolith. That broad umbrella includes a variety of sectors. For industries like coal, regulation of climate change posed an existential threat, but for others, notably insurance, it was climate change that posed the direct threat. In other words, some businesses should have had a different cost-benefit analysis than, say, a fossil fuel company.

In fact, the insurance sector did have a very different posture toward climate change than the fossil fuel sector. But when it came down to acting on that very different analysis, it was, with a few exceptions, mostly business as usual for much of the era. Warren Buffett once remarked that before the attacks of 9/11, the insurance industry assumed the risk of terrorism for free. The same could be said about climate change. Moreover, unlike prior risk mitigations such as seat belts and lighting standards, where the insurance industry was quite aggressive in lobbying Congress for action, in the case of climate change, the lobbyists were largely AWOL. It was a case of a dog that barked but didn't bite.

It's the actions of the insurance companies, a sector that was threatened by climate change itself, that provide the most telling clue to why the world dithered for decades about the problem. Throughout the climate change era, at the level of daily business, almost all of American finance and industry acted as though climate change was not an issue. This applied even to those businesses that found their profits most at risk from climate change and recognized that threat early on.

Consequently, the collective voice of the business community that the public was hearing in the early decades of this era was that climate change was not a problem, or that it was a problem that lay far off in the future, or that efforts to deal with it would cost you your job, or that climate change would make the world a more pleasant place. This voice was amplified through ad campaigns, through astroturf (fake grassroots) organizations and through paid "experts." Separately, this message was delivered to politicians by lobbying groups and political action groups. The campaign provided a case study in how to muddy public opinion and divert political action.

Nor were Americans the only people listening. Because fossil fuel interests recruited three American presidents (Bush I and II and Trump) and many other prominent politicians to their cause, China and other emerging economies heard wildly seesawing messages from the West when they were choosing their paths to development. They were being told to ignore their domestic supplies of coal and other fossil fuels and, instead, to power their development with renewables, while at the same time powerful politicians in the West were belittling global warming and championing the use of coal in their own countries.

Throughout my career, I've been drawn to explore revealing anomalies, curiosities that provide a gateway to understanding fundamental dynamics underlying an event. In the case of climate change, the revealing anomaly is that insurers aggressively pursued writing policies for projects that would contribute to global warming, and for homeowners and businesses in areas at risk because of climate change, for thirty years after they first identified the problem.

One would expect that a sector whose business is understanding risk would factor this new risk into its pricing. That insurance did not do so at the retail level speaks to a fragility at the very heart of our consumer society. In particular, this peculiar anomaly offers a window into the perverse incentives that delayed or emasculated action on climate change for more than three decades and that still pose the biggest impediment to action going forward.

The property and casualty insurance business, for instance, has long recognized that it stands right in the crosshairs of virtually every threat posed by climate change, ranging from more intense windstorms and hurricanes, to the increased risk of wildfires, to floods and droughts, to sea level rise. The industry recognized the risk almost from the get-go. If, in a changing climate, past data can no longer be relied upon to price future risks, the industry might ruinously underprice property insurance. With this in mind, many of the big reinsurance companies (those giant firms that underwrite the risk of catastrophic events) have put together conferences, reports, and studies of various risks from the beginning of the climate change era.

I've been a consumer of these studies and reports and helped edit and write one in 2004. Moreover, the industry is huge. With $640 billion in premiums each year in the United States and more than $1.6 trillion worldwide, the property and casualty insurance industry rivals the oil industry in terms of financial heft. It could have lobbied for Congress to take action on fossil fuels just as it successfully lobbied for electrical standards in wiring and seat belt laws. Indeed, when I first wrote about the insurance industry and climate change for *Time* in 1994, that was the hope—that insurers would turn out to be the white knights of the business community.

Instead, the industry has turned out to be a very timid knight. Why that is so helps us understand the nature of the inherent vice that has left society blind to the threat of climate change even as we can see climate changing before our eyes.

While the big reinsurers have done admirable work trying to model the risks, at the other end of the business—where the policies are sold—the actions of insurers have more enabled climate change than adapted to or reduced the threat. Finally, belatedly, change is happening: in July 2018, Swiss Re, the second-largest reinsurer, made headlines when it announced that it would no longer insure businesses that derived more than 30 percent of their power from coal, joining a host of other insurers that had recently made such commitments (another giant, Lloyd's, made a similar announcement in 2020). These actions make it all but impossible to finance new coal-fired plants in the West.

Coal is by far the worst contributor of greenhouse gases, producing 2.86 tons of carbon dioxide for each ton of the fuel burned (per the U.S. Energy Information Administration), so reducing the use of coal is essential. Reinsurers began denying coverage to coal-fired utilities only in the past few years. That means they've been enabling the financing of coal plants for the prior decades of the climate change era. Which part of the reinsurance industry has had more impact on the world: the modelers and strategists who produced the many great reports that have come out of the industry? Or the executives who green-lighted reinsurance coverage for thirty years for the planet's worst climate actors?

Then there is the extreme other end of the insurance business, the agents who sell the policies to homeowners. Companies like Allstate and State Farm were writing policies for homeowners in fire-prone areas of California right up until the 2018 Camp Fire cost $12.5 billion in insured losses. Then they started canceling policies in at-risk areas and stopped writing new ones. One of the state's insurance lobbyists was quoted as saying that insurers were "scrambling" to understand the new liability.

In fact, the increased risk of wildfires has been in virtually every insurance industry report on climate change I've read over the past decades. That insurers have been surprised by a risk that they've worried about for thirty years points to something else entirely. That insurers would continue to write fire coverage right up to the point where losses overwhelm premiums points to perverse incentives deeply embedded in the way we (the United States and many other developed countries) practice capitalism.

Later chapters in the book will explore exactly how these incentives played out in the case of property insurance (depressingly, all players were doing what they could have been predicted to do given the competing pressures on them at any given point), but the fact that a business predicated on accurate risk assessment should profess to be blindsided by events its own risk assessors predicted underscores the power of these incentives.

Some of the perverse incentives come from well-intentioned government programs. Had the insurers operated in a vacuum, they might have forced action on climate change. Most policies are renewable every year, and once the risk hit, the insurers could simply pull out of an area if they weren't allowed to price in the newly discovered risk. But insurers don't operate in a vacuum; they are a regulated industry. In California, regulators imposed a moratorium on cancellations after recent wildfires. And in cases where insurers still pulled out, such as Florida, where the state would not allow insurers to price for the rising risk of hurricanes, the state has provided a backstop. Across the country, various state and federal programs provide similar backstops for natural disasters such as floods, wind, and fires, all of which subsidize risks of climate change that would otherwise be expressed in increased insurance costs, falling home prices, and other economic signals.

The combination of complacency and perverse incentives is a recipe for unhappy endings. For a case study of how perverse incentives can create a financial crash, we can turn to the relatively recent near-death experience of the financial sector during the Great Recession.

In 2008, a large number of people, in and out of the financial sphere, saw a crash coming but were ignored because business as usual was just too profitable for too many people and financial institutions. To set the context, recall that one of the triggers of the 2008 meltdown was the earlier parabolic rise of home prices as the explosive growth of a new type of bond incentivized lenders to write mortgages regardless of whether a household had the wherewithal to keep up payments. The bankers got fees for writing mortgages but didn't worry about repayment, because they quickly sold the mortgages to other bankers who would use the mortgages as the collateral for billion-dollar securities. When the mortgage defaults inevitably arrived, many of these securities became worthless, and with investors pulling back, easy money disappeared from the housing market. Plummeting housing prices fueled a banking crisis that morphed into a financial crisis and severe recession, devastating the markets and throwing millions of people out of work and into bankruptcy.

One of the accelerants of the downturn involved panic selling of homes in overpriced markets such as Las Vegas, Phoenix, and parts of Florida as sellers found themselves competing with banks dumping foreclosed properties as well as forced selling by neighbors even as the pool of buyers dried up. Later chapters will explore how climate change might well bring about a similar cascade of financial repercussions.

There are many other examples of well-predicted disasters happening because those in a position to make decisions found it more profitable to ignore a certain threat with an unpredictable timeline than to sacrifice some gains for long-term security. The matrix of competitive pressures on today's businesses, both industrial and financial, incentivizes executives to drive off cliffs. Those careering toward the cliff keep the pedal to the metal because it's almost never clear how far away the cliff lies.

This tendency seems to be in the very DNA of modern American capitalism. If immutable, this does not bode well given that climate is changing around us, and those changes will continue to accelerate. On the other hand, systems mutate just as organisms do. Chinese communism is not the communism of Mao. The regulated/deregulated tendencies of American capitalism have been more of a sine wave than a straight line over the decades as politicians and the public reacted to crashes, booms, and depressions. And today, more and more businesses see climate change as a bigger threat to their future than regulation. In this lies hope.

What we don't have is time. The global community has squandered decades and even now struggles to address the issue. The following chapters show how we lost our way.

FIRE *and* FLOOD

One

THE FOUR CLOCKS

O ur present perilous situation is the product of the interaction of four different realms: reality, the scientific world, public opinion, and the world of business and finance. It's useful to imagine four clocks running at different speeds, with each clock representing one of these four realms, the latter three all lagging behind the reality of climate change, but to different degrees. In the scientific realm, there is a lag built into the very structure of scientific inquiry. Science proceeds by gathering data, analyzing it, and publishing the results; from the 1980s on, we see a built-in lag of at least two years between what has actually been happening with the climate and our scientific understanding of it. The realms of public opinion and finance have lagged even further behind reality.

Let's start all the clocks in 1979, the year President Jimmy Carter's blue-ribbon panel presented him with their recommendations.

Clock One follows the progress of climate change itself. Hottest years in history began accumulating in the mid-1980s, and since then each decade has set new records. For instance, the 1980s counted six of its ten years as among the top ten, with 1988 breaking the all-time record for global warmth. Since 2000, every year but one has been among the top ten globally, with the warming accelerating as we approach 2023 (each of the last seven years of the 2010–2020 decade was one of the seven hottest years on record). There have been many other signals of climate change, such as a rapid acceleration of sea level rise. This clock shows that human-caused climate change has been with us since the 1980s and has been becoming ever more intense.

Clock Two marks the progress of science. Despite the warnings of the Carter blue-ribbon panel, until the mid-1990s, most scientists felt that global warming relating to greenhouse gas emissions wouldn't arrive until the next century and would make its presence felt only incrementally. In part this was because, back then, those reconstructing past climates literally could not see the rapid, violent global changes that have characterized climate change since the ice ages began some 2.7 million years ago. Early tools for reconstructing global climate in the distant past had a resolution of hundreds of years, but scientists steadily improved various proxies that permitted them to look at ancient climates so that the picture of past climates emerged with ever more precision. What they saw was truly alarming and completely different from the conventional wisdom of earlier decades.

By the mid-1990s, signals of extremely rapid changes in the past had become blindingly clear, and by 2003, climate science had undergone a complete paradigm shift: where once climate change was seen as stately and incremental, the consensus now was that it could be dramatic and swift. As climate scientist Richard Alley put it, the old view was that climate change was a dial; the new view, a switch.

Speaking to the recklessness of not controlling greenhouse gas emissions given this new view of climate, Wallace Broecker, the pioneer of the role oceans play in climate change, was even more colorful: "Climate is an angry beast," he said, "and we are poking it with sticks."

Clock Three marks the progress of public appreciation of the threat. At times this clock has advanced rapidly, and at other times it has run backward. There were periods in the 1980s and 1990s when public concern spiked. Regardless, awareness of how climate changes and whether climate is changing lags behind, by a large margin, both reality and the progress of our scientific understanding. If the first clock reflects what is happening currently, and the second clock lags by two years, the third clock has lagged the first two by as much as decades. Right now, opinion is changing rapidly, but as recently as a year ago, significant numbers of Americans still resisted the idea that humans were changing the climate—something that became scientific consensus more than twenty-five years ago—and an even larger percentage were unaware that the scientific consensus has shifted to the understanding that climate changes can be dramatic and extremely rapid.

Clock Four marks the understanding of climate change in the world of business and finance, including the economics community, the markets, and investors. With a couple of exceptions, this clock lags even the public in terms of appreciation of the threat. Right now, this is changing rapidly, but until a couple of years ago, the principal way climate change captured the attention of business, economists, and investors was not how climate change might affect the economy, but rather how attempts to limit fossil fuel emissions might lower profits.

This fourth clock is perhaps the least well examined and least well understood of the causes of our present dilemma, but it is also the most important. If the markets had the incentives and penalties to price in the likely future costs of climate change, the world would have acted decades ago, and we might have forestalled the changes we are seeing today. But the market did not have those incentives, and so the markets and the business community focused on the costs of changing business as usual. To justify this posture, they had to ignore increasingly clear signals from reality, as well as a scientific consensus about the dangers. That they were able to ignore these signals as long as they did turned out to be immensely consequential.

Because the markets are where the money is, politicians followed, and we are left with our present situation where the clock of climate reality ticks in real time, the scientists are mid-decade, the public is in the 1990s, and, until very recently, the financial community remained stuck in the 1980s.

OK, then; we have four clocks, each one running at a different pace. Clock One represents reality. The lag in the scientific clock, number two, is built into the very structure of scientific inquiry and that's not going to change. The pace of the public clock, number three, varies according to events in the timelines of Clocks One, Two, and Four (business and finance), with the weightings of these various inputs also varying over time. In this sense the public clock is passive, dependent on outside influences. For most of the last three decades, events in the real world and scientific community have been less influential on the public than the messages coming from business and finance.

Today, the weightings of the influences on public awareness are shifting rapidly, as evidenced by the swiftly rising alarm in the public and a growing realization in the financial community that the reality of climate change might pose more of a risk to the economy than attempts at regulation. We will dive into the current state of play with regard to climate change in later chapters, but first we need to understand how we got here.

PART ONE

Setting the Stage

Two

THE BIGGEST
PICTURE

Were it not for the human hand, the climate would likely still be in the sweet spot (at least for humans) that it has been in for most of the past 11,500 years. It's not a coincidence that almost all of human civilization developed during this hiatus from the ice ages, nor is it coincidence that our numbers exploded during this period, rising from about 5 million to more than 7.8 billion as of this writing.

This sweet spot is a geophysical reality. Innumerable factors create a given climate, but dominant among them is where earth is positioned in space relative to the sun and how it is tilted relative to our star. Earth's orbit around the sun changes in a 100,000-year pulse. We're presently in the rounder part of this pulse, which means that earth is now at the beginning of a 100,000-year ice age cycle (note that any return to ice age conditions could be thousands of years in the future). The home planet's spin axis is now tilted about 23.5 degrees. This should accentuate the difference of the seasons, but countering that effect is where we are in the precession of that spin axis.

Precession is easiest to visualize if you imagine a rod driven through earth's poles with a pen on each end. Precession would be the circle that pen draws as the spin axis shifts back and forth on a regular basis. Right now (and for the next many thousands of years, because precession also has a 100,000-year cycle), the Northern Hemisphere is tilted away from the sun when the earth is closest to the sun, and toward the sun when it is farthest. The net effect is to reduce summer-winter differences in the Northern Hemisphere, where most of the world's food is grown. That's a good thing.

Humanity has been the beneficiary of other orbital dynamics. There's the 100,000-year oscillation of earth's orbit from rounder to flatter, the 41,000-year cycle that characterizes changes in the tilt of earth's axis, and the circle described every 26,000 years by the precession of that axis. Geophysicist Richard Alley notes that we would have to go back 115,000 years to find an equally warm, human-friendly set of orbital dynamics as we have enjoyed over the past ten thousand years.

There are also climate cycles related to natural events here on earth, cycles with periods ranging from millions of years to the quarterly changes of the seasons. Roughly 2.7 million years ago, for instance, the Panama land bridge rose, separating the Atlantic and Pacific oceans and diverting equatorial currents. Along with some other events, this set in motion changes in how heat was distributed around the world, and, voilà, the ice ages began.*

Other, shorter cycles also derived from events here on the planet.

---

* As an aside, there's a cosmic irony here because the ice ages gave a kick start to human evolution. Richard Potts of the Smithsonian Institution has argued that during the periods of climate upheaval during the ice ages, specialist mammals, including many of our hominin ancestors, tended to die out, while the bigger-brained generalists tended to survive. Potts and others trace this pattern right down to the emergence of the biggest-brained hominid of all, *Homo sapiens.*

During glacial periods, so-called Heinrich events drove down temperatures every 10,000 years or so (Heinrich events involve mass discharges of icebergs into the North Atlantic, whose melting then shuts down the currents that deliver significant amounts of heat to the northern latitudes). Another regular cycle of abrupt warming and then cooling recurred every 6,100 years (also described by Heinrich, this cycle involves a lagging response of the ice sheets to changes in solar radiation). There are many other cycles related to sunspots or to the interactions between the oceans and the atmosphere, including the now familiar El Niño/La Niña cycle that has a period of just a few years.

There have been blips over the past 11,500 years, and those blips give a foretaste of what happens when climate goes haywire. The most recent blip was the Little Ice Age. It began around AD 1300 and was at its most intense between 1645 and 1715. The Little Ice Age ended in the mid-nineteenth century. During the Medieval Warm Period that preceded the Little Ice Age, populations exploded, with England's population tripling in the 1200s. Life spans lengthened as well.

Then around 1300, climate began to whipsaw, with deep freezes alternating with hot years, droughts with floods, and Europe was battered by epic storms. Waterlogged fields became incubators for molds, pests, and disease. One of these, St. Anthony's fire, emerged from a blight that blackened kernels of rye. As described by pioneering climate historian H. H. Lamb, whole villages would succumb to convulsions, hallucinations, and gangrene. Bad as it was, St. Anthony's fire was just a warm-up act for the Black Death, which blitzed through the compromised immune systems of people weakened by famines and prior disease. The Little Ice Age stalled population growth in Europe for four hundred years, and by its deepest part in 1715, it had shortened average life spans and even shortened the people.

After that cold spell, population growth resumed its upward march. From about 1.25 billion people in 1860, population has increased more than sixfold. The spread of sanitation, the discovery of antibiotics, and huge advances in plant breeding helped foster that rise. A largely unacknowledged factor, however, was that during most of that period, at least until the 1980s, the weather offered a clement context for human advance. The lesson of history is that climate is consequential.

Apart from these regular cycles, unpredictable external shocks such as asteroid strikes and volcanic eruptions have had outsized impacts on the climate in the past. In April 1815, for instance, Mount Tambora in Indonesia exploded in one of the most violent eruptions in recorded history. As its ash and gases circled the globe, it cooled the climate in what came to be regarded as the "year without a summer." No external event in the past 150 years has had any lasting impact on climate. The eruption of Mount Pinatubo in 1991 gave the United States a cool summer in 1992, but its effects faded within a year.

After about 150 years of stability, nested in 11,500 years of stability, climate is now changing. Over the past three decades, the changes have been coming faster, and the amplitude of the changes is more extreme. What's changing climate today could be called an internal shock or, to be more apt, a self-inflicted wound. Until the Industrial Revolution, changes in the carbon balance of the atmosphere were reactions to changes in climate. For instance, as an ice age ended, vegetation would flourish, and decay would put more CO_2 in the atmosphere, which, in turn, would further enhance the warming. With our billions of tons of emissions, humanity has turned this pattern on its head, with greenhouse gases driving climate change rather than the reverse.

We began our planetary-scale science experiment with our own climate in the nineteenth century, when the coal-fired "dark satanic mills" of the Industrial Revolution began a steep ramp-up of the amount of greenhouse gases we poured into the atmosphere each year.

What scientists knew about climate change and when they knew it will be discussed in a separate section. What became evident in the 1980s was that, after about a century of increasing fossil fuel use and increasing human numbers, climate began warming noticeably. Other changes began to become manifest. A "five-hundred-year" flood hit the U.S. Midwest. In 1994, northern India suffered a heat wave of the century with ninety consecutive days of temperatures above 100 degrees Fahrenheit. Subsequently, India has been hit by many worse heat waves, including one in 2019 that led to speculation that parts of the country were becoming too hot for human habitation.

Then Antarctica started changing, first on its outermost fringes. The Larsen Ice Shelf, the continent's northernmost and thereby most sensitive to warming, began shedding huge portions of its ice. In 1995, the Larsen A Ice Shelf disintegrated, shrinking by 2,000 square kilometers (followed by the Larsen B in 2002 and the Wilkins in 2008). In the United States, floods of the century were becoming an annual occurrence, another trend that has accelerated in the new millennium.

In April and May 2011, the Mississippi River received so much water from snowmelt and a series of four storms in the Midwest that the federal government was forced to open the Morganza Spillway for the first time in thirty-seven years, deliberately flooding parts of Louisiana in order to save Baton Rouge and New Orleans. At the same time, farther west, Texas was suffering one of the most intense droughts in its history. This led to a bizarre situation in which some landowners were suffering drought on the western part of their property and floods toward the east.

The new millennium added wildfires to the growing mix of ugly surprises and extreme events. As the climate warmed, the normal winter rain bands shifted north in the American West, leading to a succession of droughts, turning large parts of the West into kindling. The warmer weather led to an explosion of bark beetles and other pests that caused mass die-offs in forests, adding more fuel waiting for a spark.

The sparks were not long in coming. Devastating wildfires swept through western states in 2012, 2015, 2016, 2017, 2018, 2019, 2020, and 2021. Amid severe drought and the hottest year in Australia's history, the worst wildfires on the planet hit the continent's southeast in 2019, burning more than seven times the area of the California fires of 2018 and killing an estimated half-billion animals. On the other side of the globe, another enormous fire raged out of control for weeks in Siberia. Then, in 2020, the 4.4 million acres burned in California amounted to double the record set just two years earlier.

While many different factors govern earth's climate, the one that matters to us right now is our own behavior as a species. While humans are powerless to change the big cycles that affect climate, these man-made impacts remain entirely within our control. How long that remains the case is currently unknown.

Reality is the clock against which all the other clocks will be measured, the Greenwich Mean Time of climate change. What has happened in reality, however, is only meaningful to the public and financial community after the changes have been interpreted by the scientific community. Are the changes naturally occurring or a response to human actions? Is climate likely to continue to change, and if so, how fast? What are the likely impacts? These and myriad other questions flow from the events.

Because the scientific community is the intermediary between changing climate and the broader public, what various scientific disciplines understood about climate, when they came to that understanding, and their confidence in their assertions were critical to the response of the public and financial communities to the changes we began seeing in the 1980s.

As the changes began to become evident, people wanted answers as to what they meant. Scientists, however, found themselves in the impossible situation of having to invent means to document and interpret the changes in climate even as the changes were accelerating. Because researchers had to design experiments and studies, collect data, analyze data, and then publish it, the state of scientific knowledge always lagged what was actually happening by a couple of years. To put this another way, to some degree climate science is condemned to live in the past, even as the present is changing rapidly. That is the nature of science. It is to the story of the science that we now turn.

Three

SCIENCE:
The Dawn of the Modern
Climate Change Era

In 1979, President Jimmy Carter was presented with the findings of an ad hoc committee convened by the National Academy of Sciences to investigate whether human activities might change climate in harmful ways. The group was led by Jule Charney, a pioneering atmospheric scientist at the Massachusetts Institute of Technology, and the committee included some of the most distinguished researchers of the day. The Charney Report predicted that a doubling of CO_2 in the atmosphere would lead to a rise in global temperatures of about 3 degrees Celsius. As of today, this prediction is right on schedule. The report also predicted that the world would see noticeable changes in climate by the end of the millennium. This was a radical prediction for its day, but it turned out to be vastly too conservative; noticeable changes would begin in the next decade.

It's important to note how courageous the report's predictions were given the state of climate science in the late 1970s. At that time, the conventional wisdom was that climate changed at a stately pace, measuring in the thousands of years. Scientists had numerous proxies to reconstruct recent and regional climate changes through tree rings and lake and seabed sediments, but as they tried to probe further back in time, the resolution of the proxies became more blurry and the proxies themselves more ambiguous. At this point, few scientists were even looking for signs of rapid climate change. Some visionary oceanographers and geophysicists argued that past climate epochs had seen very rapid changes, but in 1979, that view was an outlier. Yet here was a distinguished panel arguing that climate might change rapidly, and in just a few years. But nobody was listening.

The lack of proxies, the lack of theory about how climate might change rapidly, and an indifferent public were just a few obstacles facing climate scientists. There were very few research stations gathering data about the vast stretches of permafrost in the Far North. Researchers knew that the frozen ground trapped truly enormous amounts of greenhouse gases such as carbon dioxide and methane, but conventional wisdom was that this icy layer, which had been stable throughout modern history, would continue to be so. In any event, at the predawn of the climate change era, if warming came to the permafrost, scientists would have a hard time noticing it.

The polar regions had been much more actively studied than the permafrost, and in 1980, Antarctica's vulnerable ice shelves had been in place for hundreds if not thousands of years, and the great ice sheets looked immutable. If anything, conventional wisdom tilted toward ice sheet thickening in the coming years.

The Charney Report never addressed whether sea levels might rise. The varied impacts of climate change were beyond the group's mandate. The members had major hurdles to overcome simply to convince politicians and policymakers that humans might actually alter temperatures. A discussion of impacts could come later. At that point, sea level had been close to stable for thousands of years.

Nor were scientists noticing much change in the midlatitude and tropical glaciers of the world. Equatorial glaciers are particularly hard to maintain given the intense solar radiation they endure. Still, perhaps the most iconic home of a tropical glacier, Mount Kilimanjaro, seemed stable in 1980. It had more than a dozen glaciers then, and its ice cover dated back more than eleven thousand years.

Kilimanjaro is unique. A massive three-coned volcano that erupts from the African plain, it is the tallest freestanding mountain on the planet. Its ice fields had begun shrinking before 1980, but this was because of a shift in precipitation patterns in the late nineteenth century. Precipitation is crucial for equatorial glaciers. Not only does it replace ice lost to the impacts of direct sunlight, but a snow layer lying on top of a glacier also protects the underlying ice by reflecting the sun's rays.

In 1980, researchers knew that a number of glaciers were retreating, but the widely shared assumption was that this was due to regular climate cycles; researchers trying to predict future conditions did not expect a major glacial retreat. This meant that in those early days of trying to imagine how greenhouse gas emissions might impact lives, there was not much concern that glaciers would be a major thing to consider.

Perhaps the most important factor influencing scientific conventional wisdom in the late 1970s was that temperatures had begun to rise only a few years earlier, after more than thirty years of remaining relatively cool. That temporary cooling had to do with a number of factors, including the massive amounts of particulates thrown into the air by coal burning and other emissions in the years before clean-air regulations and other pollution controls.

The cooling preceding the 1970s produced some mixed messages. The attention to particulates and questions about where the planet stood in terms of orbital dynamics related to ice ages led a few scientists to speculate that the earth might enter a period of cooling rather than warming. In 1975, the National Academy of Sciences issued a report that mentioned this possibility but also argued that the weight of evidence favored warming.

The idea that the future held global cooling was never more than a fringe concept. A study of seventy-one peer-reviewed papers during that period by the American Meteorological Society revealed that forty-four came down on the side of warming, twenty were neutral, and seven predicted the possibility of cooling.

Despite this, the mainstream media made it seem as though a scientific consensus had solidified around the cooling hypothesis, most notably in a 1975 *Newsweek* science story. This was never the case, and by the late 1970s, the cooling hypothesis had faded completely. Scientists were now clear that if humans had the power to change the chemical balance of the atmosphere through greenhouse gas emissions, the result would be warming. By 1976, the late Stephen Schneider, the most prominent scientist associated with the cooling hypothesis, had firmly thrown in his lot with the warming camp.

While the global cooling dustup came and went rather quickly, a more persistent line of dissent came from scientists who argued that any warming resulting from greenhouse gases would be offset by changes in cloud cover as the earth warmed. Clouds do have a big impact on surface temperatures, and their response to increased CO_2 proved very difficult to model. Some clouds have a cooling effect by reflecting sunlight and providing shade, while others warm the surface by acting like a blanket. Richard Lindzen, an MIT professor, put forward a theory that warming tropical oceans would thin cirrus cover in the tropics, allowing heat to escape and thereby keep global temperatures in balance. Lindzen still promotes this so-called iris hypothesis, which has made him wildly popular with climate denialists.*

The Charney Report was a major milestone noting the threat of global warming, but it was not the first time that potential human-caused climate change received presidential attention. In 1965, President Lyndon Johnson spoke of humans running "a geophysical experiment" with the atmosphere. Prior to the Charney recommendations, a *Global 2000* report commissioned by Jimmy Carter in 1977 also referenced the problem. George Woodwell, the founder of the Woods Hole Research Center (recently renamed the Woodwell Climate Research Center), tried to get the threat of CO_2 on international conference agendas as far back as 1970. He told me that he failed because other scientists controlling the agendas insisted that there was no evidence greenhouse gases were having any effect.

---

* Regardless of whether cirrus clouds are thinning in the tropics, steadily rising global temperatures in the subsequent decades have disproven that this mechanism would keep temperatures in balance. Moreover, Lindzen's prior record as a contrarian offers a context to judge his credibility. Long before he was telling Congress not to worry about global warming, he testified before committees that there was no connection between smoking and lung cancer.

While scientists and policymakers struggled to get a grip on the issue of greenhouse warming in the 1960s and 1970s, an invaluable scientific advance on climate change came out of the most prosaic and unexciting type of science: monitoring. Though it may seem like drudgework, monitoring has been foundational, providing hard data to mobilize action on many environmental issues. Had British scientist Joseph Farman not seen anomalous readings about ozone in the upper atmosphere over Antarctica, it might have been years after 1982 that scientists discovered the ozone hole, and the delay might have been disastrous.

With regard to greenhouse gases, the decision to collect a long-term record of CO_2 in the atmosphere grew out of a distinguished oceanographer's desire to confirm a theory about CO_2 uptake by the oceans—a problem that had nothing to do with global warming. After studying the intricate chemical reactions by which oceans absorb carbon dioxide, Roger Revelle theorized that after certain thresholds were reached in the upper layers of the oceans, additional carbon dioxide would be expelled back into the atmosphere by evaporation. To confirm his hypothesis, he needed to see whether changes in the atmosphere matched his projections. It was his luck to find Charles Keeling, a Caltech postdoc student who was obsessive about such monitoring.

Gathering data, first from Antarctica and then from the top of the volcano Mauna Loa in Hawaii, Keeling began monitoring the atmosphere in 1956. The baseline for preindustrial levels of atmospheric CO_2 had been established at 280 parts per million. As early as 1958, Keeling's measurement showed levels had increased to 315 parts per million. And since then, readings from Mauna Loa have been on an unrelenting ascent, increasing to the present 420 ppm.

At many points during his long career, Keeling's funding was threatened by the deep-seated prejudice in big science that monitoring was not real science. Keeling relied on grants from the National Academy of Sciences, and George Woodwell remembers that on more than one occasion they initially denied Keeling's proposals, saying that he needed to do experiments, not "just" monitoring. Woodwell and others would intercede, saying that the work was vital, and Keeling would keep his funding.

Despite his pioneering work on the interactions of the oceans and atmosphere with regard to CO_2, Revelle regarded the threat of global warming more as a curiosity and matter of academic interest for much of his career. This proved to be fodder for denialists. In 1982, Revelle wrote a letter to *Scientific American* arguing that as of that time there was still no clear signal of warming that could be separated from natural variation, though he expected that such a signal might appear in the next ten to fifteen years. It came sooner than that.*

To be clear, no reputable scientist had issue with the basic theory of the greenhouse effect. All recognized that heat-trapping gases impacted surface temperatures. The question, in 1979, was by how much human-produced emissions would impact climate, and when those changes would occur.

One source of confusion, trumpeted by those who opposed taking action on climate change, was that even though CO_2 was increasing in the atmosphere, the amounts were tiny. Finding a concentration of

---

* Revelle's data-driven scientific caution was later seized upon by the denialists to suggest that he changed his mind about global warming. This was nothing but spin, contradicted by the public record, but it persists thirty years after Revelle's death.

315 parts per million is like finding 1 red marble in a bowl filled with 3,149 white marbles. Doubling that amount would mean that there were 2 such marbles in an overwhelming sea of white. Could that possibly matter?

It turns out it does. The relationship of CO_2 levels and global temperatures was settled long before global warming became an issue. That didn't stop deniers from raising this dissent well into the 1990s. In his book *A Moment on the Earth*, published in 1995, Gregg Easterbrook ridiculed global warming on precisely these grounds.

I chose the Charney Report as the start of the modern climate change era because it marked an inflection point in science. It contained this prophetic sentence: "A wait-and-see policy may mean waiting until it is too late." In terms of action, the world did adopt a wait-and-see policy, and we may have waited until it was too late. In the years following, rising global temperatures set off alarms, which, in turn, greatly accelerated the pace of scientific discovery.

To sum up the state of play in 1979, scientists and concerned policymakers had to deal with a distracted public and no noticeable changes in the biggest and most ominous potential impacts of a changing climate—temperatures, glaciers, the permafrost, sea level—that might have been used to get the public's attention. Other impediments to raising public concern were relatively low-resolution proxies for studying how quickly climate had changed in the past and a conventional scientific wisdom that assumed that any such changes would be far off in the future.

The authors of the Charney Report were not alone in arguing that the change might come sooner, even by the year 2000. And, at least during the Carter administration, these scientists were getting a hearing in the White House. In a memo prepared for President Carter at the urging of Gus Speth, then head of the Council on Environmental Quality (CEQ), Woodwell, with backing from science adviser Frank Press, wrote that greenhouse gas emissions would produce "a warming that will probably be conspicuous within the next twenty years." They were right, but no one was listening.

A major problem for those few scientists who recognized that climate changes might come soon was that they needed a scientific basis for any prediction of such a rapid response. A small group of scientists, notably the late oceanographer Wallace Broecker, were trying to figure out what mechanism could bring about a rapid change. Inspiring that search were teasing glimpses that climate had changed rapidly in the past. Broecker began making that argument during his days as a graduate student in the 1950s. In 1973, he published an article in *Science* entitled "Climatic Change: Are We on the Brink of a Pronounced Global Warming?," introducing a phrase that would come to describe climate change for the public going forward.*

Broecker and others who took the view that climate change was coming and that it might be rapid were marginalized in the 1970s. Broecker was known as an aggressive personality with a big ego, and this may have limited his influence at that point. "It's easy to make enemies in science," remarks Woodwell. Moreover, Broecker and others had only ambiguous evidence that such rapid changes had oc-

---

* Broecker never wanted to be considered the "father of global warming" (the paper was about a shift in natural cycles) and offered two hundred dollars to anyone who found an earlier reference to the term. In a memoir of his career published in *Geochemical Perspectives*, Broecker wrote that in his search for an earlier use of the phrase, he came upon a reference to "global warming" from 1957 attributed to an unidentified scientist. Based on the context, Broecker felt sure that this scientist was Charles Keeling.

curred in the past and, at that time, lacked a convincing theory of what might bring it about.

And yet some scientists were nonetheless willing to go on record saying that changes might be manifest in a matter of decades. Gus Speth says that such an outcome came directly from the models upon which the Carter administration's last report on the climate threat, issued in January 1981, was based. At that point, no one expected the incoming Reagan administration to act on its recommendations.

In any event, according to Speth, the overwhelming opinion of the scientists in the government was that the issue was real but that we had a lot of time to deal with it. Speth says that even Frank Press thought there were more urgent issues to address. "In the White House, there's *always* something more urgent," Speth remarked.

Even if scientific opinion had coalesced around the threat of climate change to the point where it might have galvanized public opinion, most alternatives to fossil fuels were inefficient, cumbersome, and costly. In 1977, solar cells cost about $77 per kilowatt-hour. Today, the same cell costs $0.06 per kilowatt-hour, and utility-scale solar projects generate power at $0.046 per kilowatt-hour, according to the Department of Energy. Those concerned about climate change in the 1970s were well aware that there were no cost-effective alternatives to fossil fuels. A 1977 memo to Jimmy Carter obtained by Spencer Weart of the American Institute of Physics contains just such a warning. Frank Press, the geophysicist who was Carter's science adviser, noted that "the urgency of the problem derives from our inability to shift rapidly to non-fossil-fuel sources once the climatic effects become evident not long after the year 2000; the situation could grow out of control before alternate energy sources and other remedial actions become effective."

Press was prescient, though climatic effects became evident long before 2000. We can only imagine what might have happened had Carter not lost to Reagan and the momentum of the Carter administration continued to bring down the cost of renewables. Speth, for one, is convinced that had Carter been reelected, he would have continued to push research on solar and that the day renewables achieved grid parity—the holy grail of renewables where solar, wind, and other sources compete head-to-head with fossil fuels on costs—would have come much sooner than it has.

Indeed, one thing the Carter years demonstrated was that investment in renewables could dramatically increase their efficiency. Just before Carter took office, a watt of solar power cost about seventy-seven dollars. By the time he left office in 1981, that cost had come down to less than twenty-five dollars a watt. Since then, costs have decreased by about 10 percent a year. Today, an installed watt (solar cell plus installation) costs less than seventy cents, and in some parts of the world utility-scale solar power costs less than half the price of the cheapest coal-produced electricity.

Moreover, today, almost all the new jobs in the energy sector are in renewables. They've proven their economic case. It would have made a world of difference had this happened sooner. This is one tragedy of the dawn of the climate change era.

Climate science structurally lags reality, but scientists did have some indicators of change that could be monitored. Among the most noticeable were readings of CO_2 concentrations: Keeling's findings showed that they were climbing significantly above preindustrial levels. Other indications of change would have been hard to tease from the noise as the climate had only just begun to change in 1979.* But the structural lags in science would become a greater issue in the 1980s and beyond simply because just after the turn of the decade, climate changes began coming thick and fast.

* For those interested in a deep dive into the early years of global warming, I'd strongly recommend Spencer Weart's writings for the American Institute of Physics. He's done a richly annotated, referenced, and cross-linked history of global warming—an invaluable resource.

PART TWO

THE 1980s

Four

SCIENCE IN
THE 1980s

Climate scientists entered the 1980s wondering when they might see a signal of climate change. They didn't have to wait long. The first year of the decade, 1981, was the hottest ever recorded based on reliable records, followed by another record breaker just two years later.

The 1980s were the decade in which the rise in global temperatures first began to separate itself from the noise of annual variations; it was also the decade during which scientists began focusing on what those record temperatures signified. Crucial to that understanding were investigations into an event at the end of the last ice age. Called the Younger Dryas, it was named for an Arctic flower that proliferated during its 1,300-year span and represented the last frigid paroxysm of the most recent ice age. First identified in the 1930s, the Younger Dryas followed an 1,800-year warming period that ended around 12,800 years ago. The question that bedeviled geochemists, glaciologists, and others concerned with climate dynamics was how suddenly it occurred.

After all, in the 1950s, a sudden climate change might mean a thousand years, and if that was the fastest that climate could change, society could rest easy about the threat of global warming. As early as 1960, however, some scientists, most notably Wallace Broecker, had found evidence suggesting that the onset of the Younger Dryas occurred in far less than a thousand years. The problem was that Broecker's evidence came from an analysis of deep sea and lake deposits, and no one could be sure that over the course of thousands of years those deposits hadn't been rearranged and thus provided a false record. The rearranger could be anything from sea worms to underwater landslides.

Because the question of how suddenly the shift occurred would influence how seriously the public should take the threat of climate change, what happened during the Younger Dryas became much more than a question of academic interest. Understanding what happened during the period required proxies used to reconstruct past climates, but how reliable were they? Unfortunately, at the beginning of the decade, many of the proxies were either subject to disturbance, or had insufficient resolution to capture annual changes, or were too short-term to use in reconstructing a long-term climate record. For instance, the familiar carbon-14 dating technique, which is based on the known decay rate of its radioactive material, has an error range of roughly one hundred years and so could not capture a really rapid change. Moreover, the technique is useless for dating material more than fifty thousand years old, and reconstructing climate patterns requires records that go back hundreds of thousands to millions of years.

Another complicating factor is that regional climates can change quite suddenly, an aspect of climate that has long been recognized by scientists. Tree rings and many other proxies reliably document the sudden massive droughts that may have brought down the ancient Pueblo and Mayan civilizations. The Little Ice Age too was a well-studied phenomenon, but it represented but a minor blip compared with the Younger Dryas. Moreover, in the 1980s there was still active debate about whether the Little Ice Age was a regional rather than global event (or even whether the Little Ice Age was one event or a string of several events). Those concerned with greenhouse gas emissions needed to know whether climate could undergo major shifts rapidly and whether they could occur on a global scale.

What they needed, in a word, was a paleothermometer, something that provided a precise record of annual changes going back hundreds of thousands of years, and they needed one they could trust. During the 1980s, there was a veritable renaissance in developing proxies that had much finer resolution and were subject to less distortion than the ones inherited from previous decades. By the end of the eighties, after many such breakthroughs, climate scientists had their "thermometer." It was not one, but rather a composite picture drawn from many different sources. In some cases, this meant developing entirely new proxies for past climate, while in others it entailed looking at previously collected material in an entirely new way. While scientists showed boundless ingenuity in teasing a picture of the past from cave formations, pollens, dust, ocean and lake sediments, and the remnants of long-dead shelled creatures, perhaps the most convincing evidence came from something long appreciated as a record of past climate but very difficult to study—ice.

Just as Antarctica and the summit of Mauna Loa offered ideal locations to collect data on carbon dioxide in the atmosphere, the great ice sheets of Greenland and Antarctica drew scientists looking for a long-term record of climate. The Greenland Ice Sheet is at least one million years old, and Antarctica's ice sheets are much older. What Charles Keeling was to monitoring carbon dioxide in the atmosphere, a Danish geochemist named Willi Dansgaard was to teasing out information about the past from ice cores.

Collecting and interpreting a core turned out to be a fiendishly difficult problem. You have to know where to drill the borehole (the bottom of an ice sheet can be subject to folding and other changes that distort the record of the deeper parts). Then you have to be able to drill without contaminating the cores, and you have to be able to extract a 10,000-foot-long core in such a way that the pieces fit together precisely and seamlessly. Cores toward the base of the ice sheet exist under extreme pressure, more than 150 times the pressures at the surface, enough such that air bubbles are squeezed to invisibility. Unless handled perfectly, the cores can shatter or even explode if exposed immediately to the higher temperatures and lower pressures at the surface. And then there's the problem of measuring what's contained in the cores.

Scientists had to figure out how gases moved around in the top layer of the ice sheet before they became entrapped in bubbles. They had to know how to adjust for summer snowfalls, how to read the signals that marked when one year ended and another began, and they had to ground-truth the thermometer once they had readings.

They needed to find something in an ice core that was ten thousand or a hundred thousand years old that memorialized temperatures, precipitation, and winds. They found past winds in the amount of dust blown in annually from the plateaus of central Asia. A precipitation record was entombed in the thickness of the core. As for temperature, that was where Willi Dansgaard made his most valuable contribution.

Dansgaard had been interested in the composition of water from his teenage years. At age thirty, he studied rain samples from a storm that passed over Denmark. He had studied the isotopes in water and had an idea that they could be used to reconstruct the temperatures when the raindrops had formed. An isotope is an atom with extra neutrons. For instance, oxygen normally has eight protons and eight neutrons, giving it an atomic weight of 16. In a sample of rainwater, however, one will also find oxygen atoms with nine or ten neutrons, known as oxygen-17 and oxygen-18. As temperatures fall, the air can hold less moisture, and Dansgaard reasoned that as precipitation formed and fell, the heavier oxygen isotopes would fall out—condense—first. Consequently, as temperatures fell, the remaining isotopes in the air would tend to be the lighter ones. From this it followed that if Dansgaard could calibrate the ratios of the heavy and normal oxygen atoms in air at different temperatures, air trapped in bubbles in ice would provide a record of temperatures at the point at which the bubble formed. It was forensic geochemistry that any CSI detective might admire.

After twelve years of further study and collecting samples from all over the globe, Dansgaard published his findings in 1964, providing scientists with one of the keystones for reconstructing past climates. Immediately after publication, Dansgaard dived right into applying his paleothermometer to the study of ice cores extracted from an American site called Camp Century on the Greenland Ice Sheet. Dansgaard was principally concerned with long-term shifts, but the ice core record he recovered also revealed what looked like major short-term changes right around the time of the Younger Dryas. Focused on long-term changes, the team barely paid attention to this anomaly.

Dansgaard's next major venture in ice coring took place from 1979 to 1981, again at an American installation, this one about 800 miles south of Camp Century. The enclave was part of the DEW (Distant Early Warning) Line that was developed during the Cold War to provide early warning of an attack by the Soviet Union. Neither Camp Century nor the DEW Line site was ideally situated for extracting a climate record, but Dansgaard had to make do with what was at hand. Dissatisfied with the available drill heads, Dansgaard commissioned one designed for the specific task and brought it with the team to the site, named Dye 3. The extracted ice core, more than a mile and a half long, showed the same rapid changes Dansgaard had found at Camp Century more than a decade earlier. With other investigations using other proxies finding the same indications of sudden jerks in temperatures, people began paying attention.

One of those people was Wally Broecker. In 1984, he listened to a talk by Hans Oeschger, who had partnered with Dansgaard in the Dye 3 drilling. As noted, Broecker had long argued that climate was subject to sudden jerks, not gradual change, and that these jerks might pose problems for humanity. One of the points Oeschger made was that the borehole readings showed that CO_2 changed in lockstep with temperatures. Struck by the evidence that both CO_2 and temperatures might change suddenly, Broecker started wondering what possible mechanism could bring about such shifts. Indeed, this derivative question raised by what appeared to be rapid climate changes in the paleo record turned out to be a far more difficult question than documenting the shifts themselves.

"Rapid" in the early 1980s meant a hundred to a few hundred years. Scientists confirmed these changes during the Younger Dryas through studies of lake bed sediments in Switzerland and pollen records. As various scientists turned their attention to refining the resolution of paleo records, things were coming into focus that had been easily overlooked when the assumption was that climate changed gradually. It was like a spy reading a clandestine message but not realizing that the real information was contained in a micro-dot in the period at the end of the sentence.

Over the course of the decade, climate scientists got better at see-ing the dot and also peering into it. They were discovering climate cycles that had been concealed by the poor resolution of earlier prox-ies and the self-imposed blinders entailed in the assumption that cli-mate cycles were long and stately (at one point, Broecker complained that the computer models of the time actually smoothed over results that might show sudden change).

As Broecker puzzled over what might cause the sudden changes found by Dansgaard and Oeschger, he returned to an idea he had speculated about early in his career. In 1985, he and colleagues pub-lished their theory in *Nature*. Broecker argued that only the oceans could bring about such rapid changes. He knew that enormous amounts of heat were distributed around the world by what he called the Great Ocean Conveyor. The Conveyor was an enormous river of saline water that moved through the oceans of the world and, in the North Atlantic, warmed Europe far more than other territories at equivalent latitudes. He calculated that this ocean circulation con-tributed as much heat to the North Atlantic above 45 degrees latitude as did the sun. As the pieces fell into place, Broecker offered his the-ory that periodic shutdowns of this conveyor were the mechanism that plunged the world into thousand-year deep freezes such as the Younger Dryas.

If Broecker was to argue that the Conveyor occasionally shut down, he also had to explain what might cause it. Along with other scientists, he posited that the warming at the end of the last ice age produced an enormous lake atop the vast Laurentide Ice Sheet that covered much of North America, and that when the ice sheet cracked, the lake drained in one of the great floods in human history. The flood emptied into the North Atlantic and, because fresh water is lighter than salt water, the surge left a pool of fresh water on the surface of a portion of the North Atlantic in the area where this great river within the ocean would dive, forming what's called "deep water." The lighter lid of fresh water would interrupt this flow, and without the downflowing saline water pulling ("entraining" is the word used to describe this motion) it along, the Conveyor would slow down and stop. Without warmer water being pulled northward by this massive heat pump, Europe would suddenly cool.

As temperatures fell during the Younger Dryas, sea ice spread as far south as Great Britain. The extended ice compounded the cooling as it both trapped ocean heat beneath it and reflected solar heat back into space (a phenomenon called "albedo"). Others have proposed different mechanisms for the start of the Younger Dryas (including a theorized large meteor impact), but the positing of a massive influx of fresh water remains the leading candidate, more than thirty years after the theory was published.

Also in 1987, Broecker was instrumental in launching two new studies of the Greenland Ice Sheet that would help propel the rapid climate change hypothesis toward its current role as the conventional wisdom about the nature of climate shifts since the beginning of the ice ages. What prompted Broecker's intervention was that the National Science Foundation (NSF) balked at funding Dansgaard's newly proposed ice-drilling project in Greenland since it was a European, not American, effort. Broecker convinced the head of the Lamont-Doherty Earth Observatory to pay to fly the European scientists to Boston for a sit-down with American scientists.

The idea was to mount a joint project at the thickest part of the ice sheet, in part to minimize the distortions caused by ice folding as they pulled up cores more than one hundred thousand years old. They couldn't figure out how to share ice from a single hole, but then, as Broecker told the story, Dansgaard made a dramatic suggestion. He suggested drilling two holes, one paid for by the Americans and another paid for by the Europeans. It wouldn't be a competition but a collaboration, and with two holes, the teams could confirm each other's findings. After some negotiating with the NSF, the project was funded, and the Europeans and Americans launched the two projects, separated by about 18.5 miles.

The two efforts, the American GISP2 and European GRIP, illustrate the structural lag inherent in studying climate. The Boston meeting was in January 1987. The projects launched the following year, and the drilling commenced during the following summer. Neither project reached the depths commensurate with ice formed during the Younger Dryas until the next decade, during the summer of 1993.

Once both teams saw the sharp lines that demarked the sudden changes more than eleven thousand years ago, they knew that what they were seeing would overturn the understanding of how fast climate could turn from warm to cold. Looking at the ice laid down before the dawn of human civilization, the scientists could see dramatic changes in temperature that had occurred in as little as three years.

Publication by both teams in *Nature* came shortly after the discovery—lightning speed for a peer-reviewed journal—but even then, there was much more work to be done. Were these changes unique to Greenland, were they regional, or were they global? Did other proxies show such rapid changes? What geophysical mechanism could explain them? These and other questions drove a good deal of research in the next decade.

As we've seen, at the dawn of the 1980s, the conventional scientific wisdom held that climate change was a smooth and gradual process. Throughout the decade, what were at first blurry glimpses that dramatic, rapid changes had happened in the past became ever more distinct, and by the end of the decade, most climate scientists became converts to this new paradigm of climate change. They knew immediately that if climate could undergo drastic changes in just a few years, then humanity was indeed playing a dangerous game by pumping ever-increasing amounts of CO_2 into the atmosphere.

In 1987, Broecker published a paper in *Nature* entitled "Unpleasant Surprises in the Greenhouse?" In it he explicitly warned that increasing CO_2 might produce sudden shifts in climate. Around this time, he began using his "angry beast" metaphor. Over the years, the phrase caught on, albeit within what was at first a very small world.

The scientific clock accelerated in the 1980s as institutions around the world mobilized to understand climate change. But it still lagged reality, and by the end of the decade, as hottest years ever began to accumulate, a number of researchers began to wonder whether the phenomenon of rapid climate change was already happening today even as they were studying it in the past. They also knew if that was true, science could never catch up, much less get ahead.

There were other major breakthroughs in atmospheric chemistry in the 1980s, one of which had great relevance to efforts to deal with climate change, even though it was looking at an entirely different problem. This was the discovery of the ozone hole in the stratosphere over Antarctica. It serves as a cautionary tale about how difficult it is to prompt international action on a global atmospheric issue. I will come back to this complicated dance in subsequent chapters, but here are the bare bones of the scientific timeline of the discovery.

In the mid-1970s, atmospheric chemists noticed that certain chlorine compounds, called chlorofluorocarbons, or CFCs, would break down ozone in certain circumstances. Sherwood Rowland, Mario Molina, and Ralph Cicerone showed how the release of these chemicals (which were used as refrigerants and propellants in spray cans) threatened the ozone layer, which protects life on earth from dangerous ultraviolet radiation coming from the sun.

In 1982, the story took a dramatic turn. Joe Farman, who had been monitoring ozone levels in the upper atmosphere above Antarctica since 1957, noticed something funny in his readings. Farman, like Keeling, loved the drudge work of monitoring. Unlike Keeling, he didn't have a PhD, which relegated him to even lower status in the scientific great chain of being. He was, however, meticulous. At first, he wrote off the anomalous readings to faulty measurement—Farman operated on a shoestring budget and his equipment was practically obsolete.

When he took his readings the following year, however, the drop-off in ozone levels was even more dramatic, with concentrations plummeting 50 percent. To confirm that this was not just a local phenomenon, he collected data from the skies a thousand miles from his base. NASA got interested and began investigating. When they did, they discovered to their chagrin that their data actually showed the same ozone drop-off; somehow their analysts with their billions of dollars in equipment had missed a hole in the ozone layer as big as the United States.

Between that discovery and 1987, when NASA sent a flight in to collect data on the prevalence of chlorine compounds and the destruction of ozone, a whole host of questions were raised and answered. That 1987 flight produced the data—the "smoking gun"—that settled the issue of what was causing the ozone hole.

In the 1980s, climate scientists would have loved to have had such a smoking gun. They would get the equivalent in a few years, but it would not have the metaphoric power of the ozone hole or its relatively simple solution. Even so, it took years for the world to muster the consensus to take action on CFCs, and it might not have happened but for a fortuitous set of circumstances that did not apply in the case of global warming. But as it turned out, the international effort to deal with ozone depletion would have reverberations for climate change long after the ozone issue was considered solved.

Five

THE 1980s:
A Distracted Public

There were many environmental problems on the public's mind as we entered the 1980s; climate change was not one of them. Air pollution, particularly smog, poisoned rivers and lakes, whaling, toxins, endangered species, and other near-term issues dominated the public's concerns. Climate change would not have made the top ten.

The prior decade (actually eleven years) had seen a massive surge of environmental interest, a tide so powerful that it spurred Richard Nixon, who couldn't have cared less about the issue, to solicit and sign the most sweeping array of environmental initiatives in American history. The list includes the Clean Air Act, the Clean Water Act, the Endangered Species Act, the National Environmental Policy Act, the forming of the White House Council on Environmental Quality, and the establishment of the Environmental Protection Agency (with its administrator given cabinet-level rank). The tributaries to this upsurge in concern were many: the image on June 22, 1969, of Ohio's Cuyahoga River catching fire, the obituary for Lake

Erie as a dead lake, smog, the near extinction of many species of whales, Rachel Carson's warnings about DDT in *Silent Spring,* and on and on.

The public pressure and the political response provided some important data points:

1. On the positive side: When outrage reaches a certain threshold, the environment can become a voting issue. An implied correlative: public pressure can overcome the typical politicians' fealty to special interests.

2. On the not-so-positive side: Public concern about the environment waxes and wanes, and when the public is distracted, special interests will eat away at enacted protections. This pressure is opportunistic, and its ebbs and flows tend to follow which party holds the presidency.

Consistently, Republican administrations since Reagan have, with varying amounts of intensity, pushed to weaken environmental laws and regulations. It's been said that during the Reagan years the anti-environmental rhetoric was extreme but the actions moderate, while during the George W. Bush years the rhetoric was moderate and the actions extreme. To that I will add that during the Trump years the anti-environmental rhetoric was extreme and the actions were off the charts.

One last major data point: the environmental efflorescence of the 1960s and 1970s that spurred the raft of legislation also prompted many of the best and brightest coming out of college and law school to devote themselves to environmental work. The era saw the founding of the Natural Resources Defense Council (NRDC), the Environmental Defense Fund (EDF), the Friends of the Earth, Greenpeace, and many other NGOs, as well as the repurposing of staid, conservative environmental organizations such as the Sierra Club, the Audubon Society, and the World Wildlife Fund. In subsequent years, many of these organizations formed powerful legal arms to serve as a counterweight to special interests even when the public interest was distracted.

One of the more notable young minds to attach himself to the environmental movement was James Gustave Speth (universally called Gus). Speth started out as the ultimate insider—Rhodes Scholar and Yale Law School. Immediately after getting his degree, he put his elite credentials and legal skills to work. Along with four other classmates from Yale Law School, he founded the Natural Resources Defense Council, which was dedicated to lobbying to influence environmental legislation and suing to right environmental wrongs.

Speth spent decades working within the system. There was his stint as head of the Council on Environmental Quality in the Carter White House. During the Reagan years, he helped found the World Resources Institute, chartered to work with businesses and governments to find solutions to environmental problems. Later he headed the United Nations Development Programme and still later the Yale School of Forestry (now the Yale School of the Environment).

Over the years, he moved from being an insider to what might be described as a radical insider,* going from believing that environmental problems could be solved inside the system to believing that the system *was* the problem. In expert testimony (testimony that formed the basis of his latest book, *They Knew: The US Federal*

---

* This was how Speth described his evolution during a panel discussion I participated in during the celebrations of Yale University's three-hundredth anniversary.

Government's Fifty-Year Role in Causing the Climate Crisis), he offered pro bono in a recent lawsuit over climate change, Speth wrote that the one consistent theme of administrations since climate change was recognized as a potential threat was "full throttle support for fossil fuels" despite clear knowledge of global warming and the availability of renewable energy alternatives.

His testimony came from experience. Speth took an early interest in climate change, and, in his more idealistic years, he thought government could rise to the occasion and address the issue. In 1977, as head of CEQ, he found himself in a position to do something about the problem.

Surprisingly, even though climate change was not a passionate issue for the public, it was subject to lively debate within the Carter administration. And Speth says they knew that dealing with the threat would require bringing the costs down on renewables. Carter famously installed solar panels on the White House roof, and, to support another major initiative on energy conservation, he would turn off the lights when he left a room—a gesture for which he was widely mocked.

Any administration is buffeted by competing interests, and the Carter White House was no different. Thus, while the administration fostered conservation and the development of renewables to reduce our dependence on fossil fuels, it was simultaneously advocating to *increase* their use. With sharp memories of gas lines following the Arab oil embargo, there was a big push for U.S. energy independence, and to achieve that goal, the Carter administration made a concerted effort to increase the use of coal, the very worst of the fossil fuels, for electrical generation.

There was never any doubt about where Carter's successor, Ronald

Reagan, stood on fossil fuels. He ran an explicitly anti-environmentalist campaign. Remember trees causing pollution (a forerunner of Trump's wind farms causing cancer)? As Speth put it, "He campaigned against latte drinking, Volvo-driving greenies." Environmentalists lost ground on a lot of issues during his two terms. In addition to delaying action on ozone, he slashed the budget for the EPA, cut back on enforcement, and opened up huge expanses of public land for oil and gas drilling and coal mining. And for every environmental chicken coop, he found a fox to head it.

With Democratic control of Congress and a few pro-environment, moderate Republicans in the Senate, however, there was a counterweight to his actions. Moreover, his early fire-breathing anti-environmental appointees overreached. Both James Watt (Interior) and Anne Gorsuch (EPA) were forced to resign, and Rita Lavelle, who headed the Superfund toxic waste cleanup, ended up in jail for lying to Congress.

The successors to the firebrands were far more reasonable, according to Speth. There were cutbacks on research budgets, he said, but discussions of policy on climate change proceeded. Speth also gives decent marks to Lee Thomas, who came after Anne Gorsuch at the EPA.

Congress maintained a passing interest in climate change during the Reagan years. George Woodwell testified before the Senate on the issue in 1986. He noted that without controlling greenhouse gas emissions we were likely to see severe Arctic warming, an increase in wildfires, and melting of the permafrost, which would release more greenhouse gases, creating a situation in which "warming feeds the warming." All these warnings have come to pass.

The last year of the Reagan administration saw climate change come roaring back as an issue. In 1988, a searing heat wave in the United States combined with dramatic testimony from NASA scientist James Hansen raised the profile of the issue. In a sweltering Senate chamber during July hearings organized by Senator Tim Wirth of Colorado, Hansen showed the senators his most recent model of what would happen to global temperatures if greenhouse gas emissions continued unabated. He also said that it was "ninety-nine percent certain" that the current observed warming trend was not natural but rather the result of human-sourced greenhouse gas emissions. As more than a thousand all-time heat records fell around the country, Hansen's testimony got everyone's attention.

Among them was George H. W. Bush, Reagan's vice president, who was running against Michael Dukakis to become the next commander in chief. Apart from record-setting heat, 1988 was something of an annus horribilis in terms of environmental calamities. Huge fires tore through the Amazon rainforest, biodiversity was fast disappearing as humans destroyed ecosystems from the equator to the poles (the great biologist E. O. Wilson described this crisis as "the death of birth"), and the oceans were afflicted by pollution, overfishing, and heat-related dead zones, to name just a few of the environmental issues that made headlines that year. Bush noticed and decided to run as the environmental candidate.

Inadvertently, I helped him out on this. I was at *Time* at that point and had been hired to report and write my own stories on science and environmental issues. During the summer of 1988, I was contacted by Michael Deland, then EPA administrator for New England. The conversation turned to Boston Harbor, and Deland mentioned that Dukakis, as governor of Massachusetts, had failed to avail the state of federal funds that would have paid 90 percent of the costs of cleaning up the filthy harbor, and that because the funding window had expired, the state would now have to assume the burden of those costs. He called it "the most expensive public policy mistake in the history of New England."

I thought it was worth investigating and went up to Boston. I gave Dukakis and his campaign staff every opportunity to offer an explanation, but the best that they could come up with was the bizarre assertion that Dukakis was proud of his record on Boston Harbor. *Time* published the piece in August, and the next thing you know Bush was on the deck of a boat on the fetid waters proclaiming that he would be the "environmental president." The campaign hammered Dukakis on the disgusting state of the harbor, on Willie Horton, and on that comical photo of Dukakis in a tank.

Bush also claimed that he would provide leadership on global warming. On August 31, in a campaign speech, he said, "Those who think we are powerless to do anything about the 'greenhouse effect' are forgetting about the 'White House effect.' In my first year in office, I will convene a global conference on the environment at the White House. It will include the Soviets, the Chinese.... The agenda will be clear. We will talk about global warming."

Bush actually did convene a conference once president, but sometime between his campaign speech and the conference in April, the "lobbyist effect" came into play. The Chinese were not invited, and the briefing papers for cabinet members warned them not to use the phrases "global warming" or "greenhouse effect." It's hard to see how much progress can be made in a conference on global warming if a president is too afraid to even say the words "global warming" at a conference he convened to discuss the subject. This launched what has become a Republican tradition of undercutting international efforts on climate change, with subsequent GOP presidents upping the ante even as the evidence of changing climate became more obvious.

Still, the issue did not go away, although it lost momentum once the Bush administration abandoned the field. At *Time,* managing editor Henry Muller decided to make earth the "Planet of the Year," and that "Man of the Year" issue received extraordinary attention around the world. Charles Alexander edited the issue and I was deeply involved in choosing the topics (and reported and wrote the section on biodiversity). Michael Lemonick wrote the section on climate change. In terms of newsstand sales, it was one of the most successful issues in the history of the magazine.

In sum, the 1980s were something of a mixed bag with regard to climate change. The decade witnessed the first noticeable signs of global warming. It was a decade of tremendous progress in the understanding of how climate changed (even if many of the results were not confirmed until the 1990s).

The decade also saw the first major expressions of public concern for a global threat to the atmosphere. There was both good news and bad news in this, and that requires some explanation. There are few people who have probed this issue more deeply than Anthony Leiserowitz, an expert on public opinion at Yale University.

Leiserowitz has spent decades monitoring what the American public knows about climate change and how they see the threat. One of the most enduring and notable aspects of the public's view is how uninformed most Americans have been throughout the climate change era. One of Leiserowitz's studies from 2002—fourteen years after the issue went mainstream—explored the question of the public's knowledge of what caused global warming. Among the choices were (1) burning fossil fuels; (2) population growth; (3) deforestation; (4) depletion of the ozone layer; and (5) nuclear power plants. By far the most common answer was—drumroll, please—depletion of the ozone layer! That's what 47 percent of respondents saw as the primary cause of global warming. "Fossil fuels" was the choice of only 23 percent of the people polled. Tony's survey confirmed the findings of an earlier survey by GlobeScan where Americans also chose ozone depletion as the primary cause of global warming. Keep in mind that Tony's survey came after more than a decade of a constant stream of articles and new items about climate change.

When Tony and I discussed this, I blurted out, "This is insane."

His response was immediate: "It's completely sane. It's totally rational and, of course, totally wrong."

Here's what he meant. As Leiserowitz explains it, part of the problem lies in our very structure: we humans are physically and psychologically oriented to focus on things in front of us and on the ground. "We're between four and seven feet tall, and our eyes face forward, rather than being on the top of our heads." We naturally tend toward worrying about what is in front of us, not about the threat of invisible gases miles above our heads. In fact, until the issue of ozone depletion burst upon the scene, most environmental concerns were experienced as local, even if they were global—smog, burning rivers, toxins.

Moreover, environmental hazards are but one claim on the average person's attention. There's the job, money, relationships, the family, sports, politics, financial crisis, geopolitical tensions, and so on. And then many environmental questions involve science; Leiserowitz notes that only about 1 in 200 Americans actually knows a scientist. Asked to name living scientists, the most common responses on one survey were Einstein and Darwin, neither of whom is living.

So, given this crowded mental playing field in the mid-1980s, along came scientists saying that an invisible gas was damaging the upper atmosphere above Antarctica. That this issue was able to penetrate the noise and gain a foothold in Americans' minds Leiserowitz attributes to genius marketing. The major piece of this genius was coining the phrase "ozone hole." This is something any nonscientist could imagine. People could also easily imagine that if there's a hole in the atmosphere, damaging UVB rays could penetrate to the ground and cause skin cancer. If there's a hole somewhere, asks Leiserowitz, what do you do? You patch it! So people could imagine the solution.

Thanks to the brilliance of the phrase, by the end of the 1980s, much of the U.S. public was aware that invisible gases could hurt life on earth by damaging the ozone layer. Then along comes another group of scientists, this time talking about other invisible gases that pose a threat to the upper atmosphere. This seemed to demand more attention span than the public was willing to allot to the upper atmosphere.

Worse, these scientists didn't have a catchy phrase like "ozone hole" but rather the somewhat vague "greenhouse effect" and the almost pleasant sounding "global warming." Even worse, you can't fix this with an easy patch by banning one group of chemicals. Rather, it seems, everything we do contributes to the problem.

Faced with these complexities, and the intense competition by other, immediate matters demanding attention, Leiserowitz says it's little wonder that many in the public squeezed the two issues together in their minds. "A hole in the atmosphere lets heat in," he remarked. "Totally rational and totally wrong."

He might have added "totally disheartening" to that list. As we will see, ignorance about the nature of the threat, ignorance about the degree of scientific consensus, about the scale of its consequences, and about when it might happen continued through the next three decades (as late as December 2019, 43 percent of Americans polled by *The Washington Post* and the Kaiser Family Foundation felt that plastic bottles and bags—the environmental crisis of that moment— were a "major" contributor to global warming), even as the impacts of climate change had begun to make themselves felt.

There was one other unfortunate reverberation of the ozone hole story. With regard to the attitudes of business and finance toward action on climate change, the 1980s were, as we shall see, even worse than a lost decade. They were the decade in which the business community perfected its weapons to stop any action on climate change.

Six

BUSINESS AND FINANCE IN THE 1980s: Creating the Denial Playbook

Looking back at the interaction of the business and finance community and climate change in the 1980s, two signals rise out of the noise. First, it was a decade during which the business community developed a playbook for questioning science and derailing regulation of atmospheric pollutants, which it would later put to use with devastating effectiveness in halting action on global warming. Second, it was the decade America lost its leadership in commercializing alternative energy sources to Europe and Japan after the Reagan administration slashed government support for wind and solar by 85 percent.

All in all, climate change barely registered as an issue for the business and financial community in the 1980s. Exxon, which later became a major promoter of climate denial and misinformation, invested in developing solar panels in the 1970s, though the motivation was not to reduce fossil fuel consumption but to diversify its energy portfolio. The business community was, however, very much concerned with the other big atmospheric threat of the 1980s—the ozone hole. Moreover, the business side of the ozone hole saga supplies the missing piece of the puzzle, explaining why it took thirteen years from the discovery of the problem to take meaningful action. It also helps explain why whole decades have passed without meaningful action on global warming.

Here's how the CFC story unfolded. By the late 1970s, the Carter administration had already banned the use of CFCs as a propellant and was moving toward phasing out their use as a refrigerant. Manufacturers of CFCs sprang into action to defend the chemicals.

The story of CFCs and the ozone hole is widely regarded as an environmental success story because the global community and industry eventually took concerted action to phase out the chemicals. This kind of success with climate change would very likely spell the end of modern civilization.

Compared with climate change, the dangers of CFCs were easy to grasp. For most people, climate change in the early 1980s was an ambiguous threat that seemed decades in the future. CFCs posed a clear and present danger. For humans, the risks included blindness, skin cancer, suppressed immune system, and damage to the spleen. A 1 percent decrease in ozone in the upper atmosphere was calculated to lead to a 2 to 3 percent increase in skin cancer. In nature, unfiltered solar radiation can damage crops and leave them more vulnerable to disease; it reduces the productivity of phytoplankton (the basis of the oceanic food chain); and just as it can cause blindness in humans, so too is it a danger to a vast array of terrestrial animals, none of which have the option of donning sunglasses. It's not much of an exaggeration to say that the destruction of the ozone layer posed a threat to life on earth and that by the late 1970s, most scientists involved in the issue knew it.*

The Carter administration took the problem seriously and would have likely banned all CFC use had Carter won a second term. Manufacturers could see the writing on the wall, and DuPont, the world's largest producer of CFCs, began research on alternatives to chlorofluorocarbons. Then an extraordinary event distracted both the Carter administration and the public. On November 4, 1979, Iranian militants stormed the U.S. embassy in Tehran, taking fifty-two hostages. As the crisis dragged on, the odds of Carter's reelection began dropping. He was forced to hole up in the White House during the crisis, unable to campaign. Worse, he had to fend off a primary challenge from Ted Kennedy, which kept Carter from directing his attention to Ronald Reagan. Savvy lobbyists for the CFC producers watched Carter flounder and began thinking that maybe there wouldn't be a ban after all.

Industry mounted a campaign. They pushed back aggressively against the scientists. Lobbyists demanded proof that the chemicals were actually harming the ozone layer. Some argued that there weren't enough CFCs being produced to matter, and even if there were enough, they wouldn't stay up in the atmosphere for long. They also used the go-to argument business uses against any regulation: eliminating CFCs would throw tens of thousands of people out of work.

In 1980, DuPont ramped up the pressure. It took the lead in organizing an unusual lobbying organization. Called the Alliance for Re-

---

* A recent article in *Science* asserted that the extinction of Neanderthals was partly attributable to the loss of the ozone layer during a reversal of the magnetic poles roughly forty-two thousand years ago.

sponsible CFC Policy, it drew members from both manufacturers and users of CFCs. These were groups that ordinarily would have had opposing agendas (on matters such as pricing, for instance), but they had a common interest in protecting the industry. The manufacturers invited the small businesses to join because they realized that this allowed them to bring pressure in every congressional district in the country. The campaign worked; support for further limits to production disappeared in Congress.

Industry stalling tactics got another boost when Reagan was elected. Two things happened in short order: Industry realized immediately that Reagan's new antiregulation regime was not going to pressure Congress to outlaw the chemicals. DuPont, which had expected a total ban if Carter was reelected, promptly halted work on producing alternatives to CFCs.

Then, in 1985, the story took another dramatic turn. Joe Farman's data, painstakingly collected and analyzed during the previous three years, was published in *Nature*, documenting the precipitous drop in ozone levels in the skies above Antarctica. Understanding the dangers of increased UVB radiation reaching the earth and oceans requires only slightly more sophistication than understanding that ingesting Lysol is bad for your health. In case anyone still didn't get it, Paul Newman, not the actor but the chief scientist for earth sciences at NASA's Goddard Space Flight Center, put it simply to *Space Daily*: "If there were no ozone layer, the Sun would sterilize Earth's surface."

Once again, industry found itself on the defensive. DuPont in particular was in a bind. The company had committed to stopping production of CFCs if "reputable evidence" showed that they were a hazard to the ozone layer. A hole in the atmosphere the size of a continent would seem to provide such reputable evidence.

As it turned out, a threat to life on earth was less of a concern for DuPont's C-suite than a threat to quarterly profits. Realizing that they had a friend in the Reagan administration, the company continued to seize any scrap of contrary information that might delay taking action.

With NASA confirming Farman's readings, DuPont's argument that CFCs didn't damage ozone was starting to look like a loser. So they fell back on other arguments, and a retrospective look at that record is damning. Over the years, ample evidence has surfaced indicating that the company was acting disingenuously in those crucial years between the discovery of the ozone hole and when the world finally took action in 1987.

For instance, a 1983 update from the National Academy of Sciences slightly lowered its estimates of ozone loss if production of CFCs remained flat (in fact, this was the year Farman discovered the ozone hole, though he did not publish the findings for two years). Robert Watson, a British atmospheric chemist, was at NASA during those years, and he recalled that DuPont spokesmen would show up at meeting after meeting, arguing over and over that the CFC market was mature. They could say this with a straight face because of a convenient by-product of the severe recession of 1982. Production of CFCs actually did drop that year, as did the production of most other industrial chemicals. Consequently, if you smoothed the data between 1980 and 1983, it looked like a flat market. What the DuPont spokesmen knew, but didn't say, was that tremendous demand for refrigerants in the developing world was going to spur rapid growth in the use of CFCs for years to come. Between 1983 and 1987, actual production increased by 7 percent a year. Yet DuPont officials publicly insisted until 1986 that they did not know that the market for CFCs was going to grow.

The other fallback argument for DuPont and the industry was that the chemicals would not live long in the atmosphere. When I interviewed chemists from the company for *Time*, they admitted that it was established in the 1970s that the chemicals remained in the atmosphere for decades. Yet in public documents as late as 1982, the company was still arguing that atmospheric CFCs were short-lived.

In 1986, DuPont realized the jig was up when the Vienna Convention established a framework for the phase-out of the chemicals. Was the company actually acknowledging its global responsibilities? Possibly, but more likely its executives realized that when CFCs were banned, DuPont would have a huge competitive advantage because of their head start on production of alternatives. Moreover, by this time, CFCs were a low-margin product, priced as a commodity. DuPont would have fatter profit margins when selling CFC alternatives.

On September 16, 1987, delegates from many nations met in Montreal to sign the protocol specifying a CFC phase-out. That same day, Newman was one of the scientists at Goddard receiving the data from the NASA flight into the vortex of the ozone hole. This produced what came to be called the "smoking gun," the data that showed the tight relationship between the prevalence of chlorine compounds and the destruction of ozone. As the evidence became overwhelming, even the Alliance for Responsible CFC Policy recognized that a phase-out was inevitable and endorsed the protocol.

Between 1978 and 1988, global production of CFCs amounted to roughly 19 billion pounds. Had the phase-out begun a decade earlier, as the Carter administration planned, the problem of ozone depletion might never have reached the proportions it did. As it stands, thirty-five years after the Montreal Protocol was signed, the ozone layer has not completely healed, and this year a new hole opened up in the layer above the Arctic.

In 1993, *Time* magazine published an article in which I explored the history of the CFC threat to the ozone layer. Here's how I ended the piece:

> The ozone story shows what can happen when the world underestimates problems. It also underscores the difficulty of imposing environmental regulations that clash with economic interests, especially in the face of scientific uncertainty. If policymakers wait until there is unarguable evidence of danger before they act, it may be too late to prevent serious environmental damage.
>
> This dilemma is now being faced on a related issue, that of carbon dioxide emissions and the global warming they could cause. Even though scientists are still debating how bad the warming trend might be, President Clinton has pledged that the U.S. will draw up a plan to get emissions of carbon dioxide back to 1990 levels by the year 2000. But will the plan, which may be opposed by utilities, automakers and a host of other business interests, make it through Congress? Corporate forces have already come up with their own version of the CFC alliance, called the Global Climate Coalition. One of the founding members: DuPont.

The Global Climate Coalition was established in 1989, the year the Montreal Protocol went into force. It not only included many of the companies from the old CFC alliance but also used the same experts (and some of these experts, notably Fred Singer, had earlier cut their teeth opposing anti-smoking regulations). A sponsor of the organization, the National Association of Manufacturers, lent it an imprimatur of respectability, but the group's hard-knuckle tactics came right out of gutter politics: sow doubt about the consensus on the science, challenge the motivations of advocates, dig up dirt on groups and individuals, and strong-arm politicians.

Those opposing action on climate change recognized from the outset that they had a far greater advantage combating efforts to halt global warming than they did with CFC regulations. Kevin Fay, the former director of the Alliance for Responsible CFC Policy, later became a director of the International Climate Change Partnership, a more moderate industry group, seeking more collaborative solutions than other scorched-earth industry organizations popping up to oppose climate action. As recounted by the World Resources Institute, Fay recognized "that climate change was a much more complex issue than ozone depletion." As he put it, "A handful of companies manufactured ozone-depleting chemicals, and there were a limited number of uses for them. Controlling carbon dioxide emissions would affect every aspect of life on earth." Moreover, in the 1980s the CFC crowd had a massive ozone hole staring them in the face, while with climate change, it was still difficult to tell whether the suite of hottest years was a natural phenomenon or a result of loading the atmosphere with greenhouse gases.

Phasing out CFC use should have been easy. Instead it took thirteen years from the time the threat was first recognized to the time when global action was initiated. And for all the hailing of this "environmental success story," we are still dealing with the impacts of ozone depletion (one of which is an alteration of atmospheric and ocean currents) thirty years after the treaty. If we have to wait for the climate change equivalent of the ozone hole before the world is galvanized into action, any treaty will be protecting the remains of an uninhabitable planet.

There was one other consequential story involving business, finance, and climate change in the 1980s: Reagan's tax cuts. Though the impact of Reagan's tax regime on renewables was scarcely noted at the time (with the press focused instead on how they would impact individuals, corporations, and the deficit), Reagan's reversal of tax incentives and disincentives that were a legacy of Carter's 1978 national energy initiative played a major role in the loss of America's leadership in developing cost-effective alternative energy sources. The tax changes also constituted a knife in the gut to the prior administration's efforts to promote energy conservation and fuel efficiency.

The late, great investor Leon Levy often remarked, "Give me control of the tax code, and I will give you any society you want." What he meant was that by manipulating incentives and penalties, the tax code was the most efficient way to shape behavior. It doesn't involve much bureaucracy, and, in contrast to regulation and state planning, it leaves it to the citizen to decide what to do. It also allows a country to shift the costs from society as a whole to the entity that imposes those costs. The huge taxes on cigarettes, for instance, both discourage smoking and provide revenue to offset the costs to society that smoking imposes in terms of health and lost productivity.

In its push for energy independence and to promote renewables, the Carter administration used both top-down regulation and the tax code as its tools. It imposed a windfall profits tax on oil producers; forced utilities to encourage conservation, not increased electricity use; forced utilities to buy excess energy (nicknamed "cogen") from industries that produced a lot of otherwise wasted energy; encouraged utilities to enter into long-term contracts with alternative energy sources such as solar and wind; and provided tax incentives for investing in solar and other alternative energy technologies.

The Reagan administration undid almost all these initiatives. The investment tax credit for solar, for instance, was killed in 1986 to fund the Reagan tax cuts. The late Philip Clapp, a major figure in the environmental movement until his death in 2008, said that Reagan "set back solar a decade."

The real cost of the 1980s was one thing that did happen and so much that didn't happen. Industry used the 1980s to optimize a playbook that would be used in the 1990s and beyond to hamstring any efforts to avert climate change. And business interests, allied with a like-minded Reagan administration, used the decade to snuff out virtually every climate-related initiative launched by the previous Carter administration.

For business and finance, the clock on climate change had barely started ticking. During the decade the moneyed interests had other fish to fry. Until the end of the decade there was no movement to control greenhouse gas emissions, and even if the scientific evidence for warming had been as dramatic as that for the ozone hole, fossil fuel interests could rest assured that they had a friend in the Reagan administration (and it turned out they had one in the Bush administrations as well). With tax incentives gone and government support for research and development on renewables cut to the bone, financial interests had little incentive to push development of alternatives. With regard to climate change, the 1980s were indeed a lost decade. Unfortunately, it was only the first.

What might have happened had Carter been reelected? He was a victim of external events, notably the Iranian hostage crisis and the primary challenge from Ted Kennedy. No Ted Kennedy and no hostage crisis, Carter might have won a second term. With continued momentum on the phase-out of CFCs, DuPont would have continued work on alternatives, and ozone depletion might not have been as severe.

Perhaps the Alliance for Responsible CFC Policy wouldn't have been as effective at slowing action. In turn, that might have led industry to be more cooperative in the next decade, rather than adversarial, as climate science and a suite of record hot years made the case for controlling greenhouse gases more compelling. With tax and other incentives still in place, costs of solar, wind, and geothermal would have come down more quickly, advancing the date when various renewables would have grid parity with fossil fuels. Today, many of the largest companies recognize the threat of global warming and resisted the Trump White House's efforts to lower fuel economy standards and otherwise promote the further release of greenhouse gases. Could this have happened thirty years earlier? And if it had, would the climate changes we are seeing be less extreme? We will never know.

PART THREE

THE 1990s

Seven

REALITY: Ominous
Portents of Change

B y the 1990s, news of new global temperature records and other weather extremes had become so frequent that, increasingly, they were treated by scientists and the media as signals that climate was already changing. The decade started off with a bang, with 1990 setting a new global record for warmth. But then in 1991, Mount Pinatubo erupted in the Philippines, throwing a huge cloud of ash into the atmosphere that circled the globe. The ash cooled the planet and interrupted the drumbeat of warming years. This precipitated a spate of "Where's your warming?" taunts from the nascent climate denial crowd.

While temperatures took a short break in the early part of the decade, a series of intense storms gave those worried about climate impacts something else to chew on. In September, Typhoon Mireille hit Japan, the first to strike the island nation in thirty years. Then, in October 1991, one of the most intense northeasters in history hit the New England coast. "The Perfect Storm," immortalized by Sebastian Junger, hovered off the coast for days, its 100-mile-per-hour winds whipping up huge waves and storm tides. Less than a year later, Hurricane Andrew hit Florida as a category 5, making landfall in Homestead, just south of Miami. Ultimately, Andrew inflicted $50 billion in property damage (in 2020 dollars) and drove eleven property and casualty insurers into insolvency. Damaging as it was, insurance experts realized that the losses could have been far worse if the landfall had just been a few miles north. The losses had a profound impact on the insurance industry.

Along with Hurricane Iniki, which hit Hawaii a month after Andrew hit Florida, these storms seemed to support a theory that MIT meteorologist Kerry Emanuel had been developing since the 1980s—that greenhouse gas emissions would spur an increase in extreme weather events, specifically that global warming would lead to more intense hurricanes. A lot of things have to fall into place to create a hurricane, but a basic ingredient is an energy source, which means ocean water temperatures above 80 degrees Fahrenheit. A warming ocean that stays warm longer expands the potential for a longer hurricane season and more intense storms.

Moreover, far more complicated interactions in a warmer globe seemed to suggest increased frequency and intensity of other weather extremes such as tornadoes, windstorms, rain events, heat waves, and droughts. In subsequent years, nature obliged by providing evidence of them all. That "five-hundred-year" flood in the Midwest caused $13 billion in damage. The following year, the American Southeast was hit by a "flood of the century." Another such flood inundated the Red River valley in the upper Midwest in 1997. That same year saw the strongest El Niño in recorded history, and estimates of worldwide damage range as high as $100 billion.

The El Niño also provided a case study of how climate change, in concert with other destabilizing factors, could bring down governments, in this case the Suharto regime in Indonesia. By the time of his fall in May 1998, Suharto had ruled Indonesia for thirty-one years. By the end, Indonesians had had it with the corruption, cronyism, and nepotism of his reign, and his government was already teetering from the repercussions of the Asian financial crisis of 1997.

What brought it all to a head, however, was the El Niño and its attendant two-year drought, which caused a tripling of rice prices by March 1997, leading to starvation in some provinces. The price hikes didn't help farmers because the price of fertilizers and pesticides rose even faster as the rupiah devalued. The El Niño also had second-order impacts, such as drought-fueled fires, which shut down airports and left seventy million people with respiratory ailments. The smoke from the fires further reduced rainfall by overloading the atmosphere with nuclei too small to allow the formation of raindrops heavy enough to fall to the ground.

By 1998, the number of Indonesians living in poverty had quadrupled to more than one hundred million people. Civil order began to break down. Suharto tried to hold on, rigging elections, but massive riots and protests in May ultimately cost him the support of his cabinet. Purportedly, he spent much of the time after his resignation tending to his pet birds and watching nature and animal shows on the Discovery Channel and Animal Planet.

The El Niño had massive impacts around the world, producing floods across huge swaths of Latin America and China, as well as droughts and outbreaks of disease in Africa. While there was blanket coverage of El Niño's impacts (everybody loves a dramatic weather story), most of the public didn't make the connection between the destruction of this event and the potential economic impacts of climate change. Perhaps this was because this El Niño was just a stronger version of a naturally occurring cycle, or perhaps it was because climate change was still viewed as theoretical and far off in the future.

Ironically, the lasting impression left by this El Niño may have been to undercut the case for global warming in the public's mind. That's because the El Niño temporarily warmed global temperatures by about 2.7 degrees Fahrenheit, setting a record for warmth that would not be broken until 2005. In a masterpiece of casuistry, those opposed to taking action seized on this anomaly to start trumpeting a "pause" in global warming. They argued that if you drew a line from 1998 forward, the temperature record looked flat. To accept this, one had to disregard that every year after 1998 was one of the top ten warmest years in the temperature record to that point.

In 1998, the United States was hit by a record-setting 1,424 tornadoes. And once the effects of the Pinatubo eruption subsided early in the decade, the upward march of temperature resumed. In all, six of the ten years in the decade were either the warmest or second-warmest years for average global temperatures.

Most ominously, signs of change emerged from the most stable place on the planet: Antarctica. In the space of a week in 1995, the Larsen A Ice Shelf disintegrated. It dumped an area of ice more than twice the size of New York City into the ocean. Before it collapsed, the shelf had been stable for hundreds of years.

The decade also gave the world a preview of a host of second-order impacts of climate change. One of the basics of evolutionary biology is that when ecosystems become disrupted, opportunistic, fast-reproducing species tend to proliferate. Pathogens fit this bill perfectly. History is replete with examples of plagues and pandemics following episodes of extreme weather.

The late Paul Epstein, a Harvard epidemiologist, pioneered the study of the relationship between climate change and disease. According to Epstein, the synergies of a warming climate and disease involve much more than reproduction. For instance, as temperatures warm, mosquito metabolism increases, and the mosquitoes respond by feeding more vigorously. The same circumstances that enhance their spread also weaken the immune systems of their victims. And once ordinarily benign microbes invade humans and animals weakened by heat and pollution, they can become deadly and invade healthy populations. Epstein noted that a class of viruses that includes measles was implicated in the deaths of seals in the North Sea, lions in the Serengeti, and horses in Australia—very different animals widely scattered around the world. In each case, said Epstein, abnormal weather had caused malnutrition, weakened immune systems, and spurred the reproduction of viruses.

Warmer temperatures also allow microbes to migrate into areas formerly too cold for them to survive. This happened in 1995, when warming temperatures allowed mosquitoes carrying dengue fever to cross the mountains that had formerly confined the disease to the Pacific Ocean side of Costa Rica. Epstein's thoughts have a chilling resonance today. The real threat, he told me at the time, may not be a single disease, but armies of emergent microbes causing havoc among a host of creatures. "The message I take home," he said, "is that diseases afflicting plants and animals can send ripples through economies and societies no less disastrous than those affecting humans."

Climate change got plenty of coverage starting in the 1990s, with many exploring the connections between various weather events and the likely repercussions of a changing climate. I wrote a good number of pieces on the subject, major articles for *Time* magazine and op-eds and essays for *The New York Times*, the *Los Angeles Times*, and other publications—frequently enough that I was described as a "global warming freakazoid" by a right-wing blog. Many others were writing about it as well.

And yet the real story of the 1990s was that basically nothing happened to slow the growth of greenhouse gases in the atmosphere. The decade ended with greenhouse gas concentrations rising to levels they would have risen to if no action was taken to contain them. In the 1980s, the global community had a number of excuses not to take action. By the turn of the new millennium almost all of those excuses had evaporated. If the 1980s were a lost decade because there was only a hazy picture of what was happening, the 1990s were another lost decade despite a spotlight beamed on the problem from start to finish. The Kyoto Protocol, 1997's limp attempt to reduce fossil fuel emissions, did not enter into force until 2005. Despite the promises following 1988, nothing happened in the 1990s to reduce dependency on fossil fuels.

Eight

CLIMATE SCIENCE
IN THE 1990s: A New
Paradigm Emerges

By the middle of the 1990s, the scientific consensus had jelled that humans were causing the climate to change. There was no ambiguity about soaring temperatures; a new and thoroughly alarming paradigm had emerged about how climate changes. If the emergent consensus in the scientific community had reached the public, perhaps there might have been meaningful action. The message did not reach the public, however, as various interests intervened to muddy and sometimes contradict the scientific message. The 1990s turned out to be a decade during which the science became clearer but the message more ambiguous.

For climate scientists, the 1990s were both a golden age and a period of tremendous upheaval. The ground shifted in the sense that a number of fundamental assumptions about climate change that held as the decade began were overturned by its end. During the same period, paleogeochemists and other researchers invented and refined tools that provided a more precise look forward as well as a new interpretation of past events, including the role of climate in human history and evolution.

It was a decade during which multiple paleoclimate investigations confirmed the hypothesis that climate could change globally with astonishing rapidity. It was a decade in which scientists began to realize that the great ice sheets were not as stable as had been assumed. It was also a decade in which scientists realized that the permafrost, which covers nearly a quarter of earth's exposed land, was not as permanent as thought. This raised the prospect of runaway global warming as thawing released truly gargantuan amounts of greenhouse gases formerly trapped in ice and frozen soil.

It was also a decade in which the public and policymakers remained largely unaware of these developments.

The reasons were manifold. The public had only recently realized that what happened in the upper atmosphere—e.g., the ozone hole—could hurt us down here on earth. To get people to believe that there was *another* threat in the upper atmosphere caused by *another* set of invisible gases was asking a lot, particularly since this threat—global warming—actually sounded kind of nice.

Another reason, however, was that the dire new findings coming from climate scientists weren't getting through to the public in a timely fashion. This was in part because the denier cohort began putting out disinformation in earnest. Another major reason, however, had to do with the way the scientific findings were disseminated.

The public does not read scientific journals. Neither do most policymakers. Rather, most people get their scientific information from the media, mainly from mainstream media, the networks, and radio. An item on the evening news might be a couple of minutes, and most articles in newspapers and the big-circulation magazines range from a few hundred to at most a couple of thousand words. This forces simplification. Worse, most mainstream press presumes (accurately) that the public has no background knowledge, so valuable space gets consumed retelling the basics of what the greenhouse effect is, leaving virtually no space or time for the actual scientific discovery. Moreover, in the interest of balance (with energetic prompting from the denial crowd), mainstream media, particularly in the 1990s, bent over backward to give space to contrarians long after the scientific consensus jelled.

With a few notable exceptions, the scientists didn't help matters. There's a cavernous gap between the demands of scientific rigor and the shorthand of mainstream journalism. Scientists hate to venture beyond the data, and most view oversimplification as a mortal sin. This makes for deadly dull and opaque quotes and confusion in the public, leading to an attitude that might be summed up as "What, you're telling me that the heat wave may or may not be related to global warming? Come back to me when you've figured it out."

If all these factors weren't enough to defang the startling and alarming scientific discoveries of the 1990s, the decade saw the advent of one other filter that was formed to collect and disseminate the latest science on climate change but, instead, effectively doused whatever flames cutting-edge science sparked: the IPCC.

That was not the intention of its original promoters; quite the contrary. The Intergovernmental Panel on Climate Change emerged as a new institution, formed to gather, interpret, and find consensus on major issues involving climate change. A creation of the United Nations Environment Programme and the World Meteorological Organization, this massive undertaking became the primary, authoritative source that the media and policymakers turned to for the state of the science. Conceived with lofty ambitions to provide policymakers with the scientific consensus on the threat and possible paths forward, what it actually did throughout the 1990s and well into the first decade of the new millennium was to understate the problem and provide ammunition to those who would delay action on climate change. The IPCC's very structure offered innumerable opportunities for mischief. Had this filter not existed between the public and the scientists, it's possible that there might have been a better chance for meaningful action on climate change in the 1990s.

The mischief came both from governments and business and finance, which had largely ignored the issue through much of the 1980s. Toward the end of that decade a number of corporate interests began to realize that they needed to do something about climate change, not to lessen the threat, but to throw gum into the works of any efforts that might arise to actually do something about it.

The case of the IPCC shows the subtlety of the business community's efforts to influence actions on climate change. The media, and thus the public, perceived the vast coalition as the voice of science, and for the most part that's what it tried to be. Because it wanted to be a big tent, however, it opened the door for various nonscientific interests to weigh in, and this in turn allowed special interests first to apply direct and indirect pressure to shape the scientific consensus, and then call into question those parts of the consensus they had not been able to influence.

These interests had their own lobbying organizations. As noted, in 1989, the veterans of the ozone battle formed the Global Climate Coalition (GCC). Given its heavyweight membership, which then included the oil majors, the U.S. Chamber of Commerce, General Motors, and a who's who of industry lobbying groups, the organization had great power in the halls of Congress. The IPCC had been formed a year earlier, and it presented industry with an opportunity to slow things down while seeming to be responsible corporate citizens.

The idea behind the IPCC was to bring together experts on all aspects of climate science, including social scientists, demographers, and others who could explore likely impacts, as well as representatives and policymakers from all nations. Their job: to assess "the scientific, technical and socioeconomic information relevant for understanding the risk of human-induced climate change."

The devil himself could not have created a more cumbersome structure containing more conflicting agendas. At that time, the developing nations regarded climate change as a rich nations' problem and rich nations' responsibility. Powerful oil-rich countries had no interest in lessening dependence on fossil fuels and every interest in undermining the science. The corporate lobbyists who wanted to thwart any momentum to act on global warming may well have thought the IPCC was a gift from God.

The IPCC was also its own worst enemy. Apart from innumerable opportunities to introduce uncertainties and round the edges of various reports, the unwieldy nature of the IPCC virtually guaranteed that its reports would be tepid and bureaucratic. For one thing, the panel included thousands of participants, all of whom could weigh in on the reports. As Gus Speth put it to me, "You've got to remember the IPCC is the *intergovernmental* panel, not the *interscientific* panel or the *interacademic* panel." Even if the IPCC had consisted entirely of scientists, the inherent caution of scientists to go beyond the data would have likely softened its findings. The IPCC, however, also included participants who actively worked to magnify uncertainties and undermine conclusions.

For instance, GCC representatives reviewing the Second Assessment Report of the IPCC took exception to some normal edits of parts of the report and used this purported "gotcha" moment to question the integrity of the process. *The Wall Street Journal* editorial page amplified their criticisms, and the deep-pocketed organization ran ads in major newspapers questioning the integrity of the IPCC process.

The government representatives didn't help matters. Having them weigh in on a process that was supposed to provide the best scientific basis for a treaty to reduce the threat of climate change may have been necessary, but the bureaucrats, politicians, and functionaries proved to be a drag anchor on the process and the reports.

Consequently, for the first twenty years of its life, the IPCC gave as much fodder to those who would delay action as to those who demanded action. Things changed in the 2010s, but by then the damage was done. The IPCC reports, at least in their Executive Summary and Summary for Policymakers sections (the only sections that most people read), consistently underestimated or understated changes such as sea level rise, the contributions of melting permafrost, the state of the cryosphere, and other areas of concern. Because the IPCC reports were taken to represent the state of the art in science, they defused any sense of urgency about the threat and gave comfort and cover to those who would delay action.

The understatements and confusion spread by these massive reports were hardly the fault of the scientists involved. They were charged with the near-impossible task of understanding fundamental aspects of an immensely complicated dynamic even as it was unfolding. They also were being called upon to make predictions about future impacts even as the basic science itself was evolving.

The actual chapters of the reports were, for the most part, devoid of political influence. It was in the opening summaries, a distillation of the major findings in each report, that nonscientific influences had some impact. Unfortunately, these summaries were taken to be the scientific consensus, when in fact the language, and what was included, suffered from the influence of political, commercial, and foreign-policy agendas.

The first IPCC assessment was published in 1990. It was a massive document, but it's a safe bet that few nonscientists read it, and those media types and policymakers who did mostly confined themselves to the Summary for Policymakers (or the press releases summarizing the Summary for Policymakers). Which means that for the public, the Summary for Policymakers became the consensus of the scientists even though it was not (the chapters themselves cautiously offered ranges of outcomes for different scenarios of climate change and its impacts).

This is an important distinction. Even as the Summary for Policymakers offered bland statements on a number of issues, the scientists themselves were finding evidence that the threat was far more imminent and dangerous than the headlines extracted from the summary. For instance, in his 1988 testimony before the Senate, James Hansen of NASA said it was 99 percent certain that the observed warming was caused by humans, not natural. Two years later, the IPCC summary held that the "observed increase could be largely due to this natural variability." It went on to hold that "unequivocal" evidence of human-caused warming would not be clear for a decade or more, though it did say that by the time the evidence was unequivocal, the world would be locked into further warming.

In the 1990 IPCC report, the chapter dealing with permafrost offers projections for how much permafrost would melt as the globe warmed, as well as specific consequences of such melting. Yet the Summary for Policymakers offers the reassuring statement that no changes in the permafrost might be expected before the year 2100. With regard to the great ice sheets, the summary noted that studies had shown changes in the ice streams within the West Antarctic Ice Sheet but that "within the next century, it is not likely that there will be a major outflow of ice from West Antarctica due directly to global warming."

All of this proved to be wrong—but not before these bland assurances took the wind out of any sense of urgency. Taken altogether, the explicit and subliminal messages of the summary were: "We've got time! No need to rush in! Much is unsettled!" Basically the opposite of a call to action. And an eager crowd of ideologues, vested interests, economists, and lobbyists were there to amplify the message.

Outside of the Executive Summary of the first assessment, science was progressing quite rapidly in the 1990s. The early part of the decade started off with a bang. Indeed, 1993 might be described as the year everything changed scientifically.

This was the year that the coordinated efforts of the European and American ice core projects published their findings in *Nature*, confirming that the Younger Dryas cooling started and ended far more abruptly than any model of past climate might have predicted. As physicist and science historian Spencer Weart put it, "How abrupt was the discovery of abrupt climate change? Many climate experts would put their finger on one moment: the day they read the 1993 report of the analysis of Greenland ice cores. Before that, almost nobody confidently believed that the climate could change massively within a decade or two; after the report, almost nobody felt sure that it could not."

Despite the fact that Wallace Broecker had been writing about rapid climate change for well over a decade, these papers could be said to have launched a revolution in the understanding of how climate changes. It was these studies that prompted Richard Alley to remark that climate change that had once been regarded as a dial now seemed more like a switch.

Dramatic as these findings were, it still took the better part of a decade for this realization to move from a startling discovery to the new paradigm for understanding how climate changes. That said, a decade is still amazingly fast for a scientific revolution (twenty-eight years later, this revolution still hasn't percolated through to most of the public). Other scientists rapidly filled out the picture.

That same year, for instance, Gerard Bond published an article in *Nature* showing that North Atlantic seabed sediment cores confirmed the signal of rapid climate change in the Greenland ice cores. A year later, Broecker, also publishing in *Nature*, proposed massive discharges of icebergs during glacial times as triggers for these rapid climate change events. The record showed that these events occurred on a millennial scale following the discharges. In honor of his colleague, he called these events "Bond cycles."

Bond followed up the next year with an article in *Science* in which he argued that the record showed that these climate flips could occur in warm periods as well as during glacial spells. Working with another colleague at Lamont-Doherty, in 1997 Bond and Peter de Menocal published an article in the American Geophysical Union's *Eos* offering evidence that abrupt shifts in climate extended into the Holocene, our present period, which previously had been thought of as extremely stable. That same year, Broecker connected these abrupt changes with disruptions of the thermohaline circulation, what he called the Great Ocean Conveyor, the global system of currents that, among other things, distributed heat to the North Atlantic, warming much of Europe. In that article, Broecker warned that if human-sourced greenhouse gas emissions triggered one of these disruptions of the ocean circulation, related atmospheric disruptions might lead to "widespread starvation." He also noted that society would find it hard to adapt to a changed climate regime because such periods are characterized by "flickers" in climate as weather whipsaws back and forth from warm to cold and wet to dry as the system seeks a new equilibrium.

These are just a few of the scientific breakthroughs that advanced through the decade. By the close of the 1990s, climate scientists knew that Wally Broecker's "angry beast" metaphor was dead-on and that past climates had been characterized by abrupt changes. They also had a far better idea of the interactions of the oceans and the atmosphere that might trigger these changes, and they knew these changes weren't smooth. A study of climate whipsaws that characterized the transitions between these events led Kendrick Taylor and his colleagues at Nevada's Desert Research Institute to coin the phrase "flickering climate."

They also knew that big trouble was brewing with the West Antarctic Ice Sheet (WAIS), even if they weren't sure when that trouble might arrive. WAIS is a marine ice sheet, which means it sits on bedrock instead of floating. Such sheets are inherently unstable. If the amount of ice they shed sufficiently exceeds ice accretion, it can set in motion a cascade of positive feedbacks that lead to the collapse of the sheet (and a complete collapse would raise global sea level by about 10 feet).

When I traveled to the WAIS in the fall of 1996, glaciologists were already seeing evidence that the daisy chain of feedbacks had begun. Ice streams—essentially rivers of ice within the ice sheet that move mass from the interior of the sheet to its edges—had already begun to flow more quickly. When I spoke with Ted Scambos, of the National Snow and Ice Data Center in Boulder, Colorado, he noted that when he looked at satellite imagery of the vast ice sheet, "I see an ice sheet in the process of collapse." At that point, however, estimates of how long that process might take varied wildly.

In the 1990s, the assumption was that the WAIS had remained roughly in its present size for 120,000 years (subsequent research has shown that the ice sheet came close to collapsing at about the time of the Younger Dryas). It seemed audacious to speculate that a century of rising global temperatures might destabilize something so massive and persistent. At the same time, scientists were documenting ominous signs. It rained at America's McMurdo Sound base for the first time that anyone could remember. The lake ice was thinning in the Dry Valleys, an area of Antarctica that was thought to have been relatively stable for more than a million years, and some of the ice shelves on the Antarctic Peninsula had begun to collapse.

These whispers of change would become a clamor over the next two decades. Even in the mid-1990s, the drumbeat of anomalies being recorded on the continent should have been sufficient to raise alarms. After all, the stakes were the potential inundation of coastal lands supporting several hundred million people.

Along with many other journalists, I wrote about these findings as they unfolded, in my case mostly for *Time*. Every one of the alarming trends uncovered in the 1990s has been subsequently confirmed; indeed, the worst-case scenarios of these years for issues ranging from sea level rise to incidence of extreme events are now either the conventional wisdom or, in some cases, the best-case scenarios of today. Still, the public remained unengaged. In part this was because there was a competing narrative on the science, one that gave plenty of cover for those who wanted business as usual to continue while looking as though they were seriously considering the issue.

The rising sense of alarm did not make it into the Executive Summary for the first IPCC assessment, and it did not make it into the Summary for Policymakers in the second IPCC assessment, published in 1995, seven years after James Hansen had told the U.S. Senate that it was certain that human influences were already affecting climate. Five years after its first assessment, the Summary for Policymakers said that there was a "discernible" human influence on climate beyond natural variation.

Updates on other climate-related threats were equally bland. The second assessment's best estimate for sea level rise was actually 25 percent lower than the best estimate offered in 1990. While the first assessment had an extensive section on permafrost (even though the Executive Summary held that no changes were expected before 2100), the second assessment barely mentioned the issue. While the second assessment gave a range of 1 to 3.5 degrees Celsius warming for various scenarios, it also held that the oceans would delay when these "equilibrium" temperatures might be reached.

The second assessment did address rapid climate change in its chapters, noting the mounting evidence that changes in the past had occurred during human time scales, but then took the wind out of the sails of any sense of alarm with this sentence: "The relevance of past abrupt events to present and future climate would be more convincing if the suggested high climate variability in the Eemian* was confirmed."

The IPCC also muffed the crucial question of tipping points—the thresholds past which processes that would accelerate sea level rise (such as the collapse of ice sheets) or accelerate climate change (such as the melting of permafrost, which would release greenhouse gases and enhance warming) might happen. During the 1990s, the IPCC reports estimated such thresholds at 5 degrees Celsius or more, a warming that looked unimaginably far off in the future back then. More recent IPCC special reports have lowered that threshold to 2 degrees Celsius, and in some cases 1.5 degrees Celsius, thresholds that are very much in the near-term future. So, back when we might have done something about climate change, the most alarming possibilities were not of immediate concern; now, when we have vastly less time to avert further temperature rises, we are told that tipping points are right around the corner.

While the publication of the Greenland ice core studies in 1993 greatly accelerated the realization that many past climate shifts were violent and sudden, the Summary for Policymakers relegated this fundamental concept to a parenthesis in its concluding section headlined "There are still many uncertainties." That section also had a one-sentence mention of a potential disruption of the thermohaline

---

* The period before the last glacial cycle from 130,000 to 115,000 years ago.

circulation in the North Atlantic, as well as the nonlinear nature of the climate system. Indeed, a look at the whole concluding section of the Summary for Policymakers underscores the unperturbed, leaden language of the document:

Many factors currently limit our ability to project and detect future climate change. In particular, to reduce uncertainties further work is needed on the following priority topics:

- estimation of future emissions and biogeochemical cycling (including sources and sinks) of greenhouse gases, aerosols and aerosol precursors and projections of future concentrations and radiative properties;
- representation of climate processes in models, especially feedbacks associated with clouds, oceans, sea ice and vegetation, in order to improve projections of rates and regional patterns of climate change;
- systematic collection of long-term instrumental and proxy observations of climate system variables (e.g., solar output, atmospheric energy balance components, hydrological cycles, ocean characteristics and ecosystem changes) for the purposes of model testing, assessment of temporal and regional variability and for detection and attribution studies.

Future unexpected, large and rapid climate system changes (as have occurred in the past) are, by their nature, difficult to predict. This implies that future climate changes may also involve "surprises." In particular these arise from the non-linear nature of the climate system. When rapidly forced, non-linear systems are especially subject to unexpected behaviour. Progress can be made by investigating non-linear processes and sub-components of the climatic system. Examples of such non-linear behavior include rapid circulation changes in the North Atlantic and feedbacks associated with terrestrial ecosystem changes.

Imagine a different ending to that summary, one that stressed that every finding since the last summary suggested that climate shifts are violent and sudden, marked by whipsawing between hot and cold years and increased extreme events; that there were ominous signs of instability in the world's largest marine ice sheet; that a substantial portion of the permafrost was in danger of melting and thereby releasing vast amounts of greenhouse gases that would in turn accelerate climate change. Imagine that this summary concluded that these indications were sufficiently dire that the world did not have the luxury of resolving scientific uncertainties before taking action to limit emissions of greenhouse gases.

That different summary could have been written—with scientific justification—but it wasn't. No politician anywhere, reading the second assessment's summary, was going to lead the charge to battle the fossil fuel lobby. Instead, the IPCC reports gave plenty of ammunition to anyone who wanted to delay action on climate change—a group that turned out to be practically everybody, ranging from the business community to the vast majority of world leaders.

There turned out to be two scientific clocks in the 1990s. It was a decade of tremendous breakthroughs in the understanding of the gears of the climate system, a decade that saw a rapid closing of the gap between the reality of climate change and our understanding of what was going on. This clock, however, was advancing much more rapidly than the scientific clock mediated by the IPCC.

Nine

THE PUBLIC:
Stirred but Not Shaken

The battle to avert the climate changes we are seeing today was lost in the 1990s. During that decade, the scientific consensus jelled, but nations dithered on any actions to avert the threat, and the big emerging nations chose a path to industrialization that locked in massive future emissions, a decision that future historians may well rank as the fateful turning point that put our civilization on a course for self-destruction. Could it have been otherwise?

Today, a number of observers say no, we could not have done anything about climate change back then. According to this logic, in the 1990s—the decade when we needed to begin to take action if we were to forestall the changes we are seeing now—there wasn't the political will necessary to begin a global shift away from fossil fuels, and even if that political willpower existed, the alternative sources of energy were too inefficient and expensive to fill the breach.

It's a reasonable argument, and if it was the case, it constitutes an utterly damning indictment of modern society. It would be saying, in effect, that the most advanced civilization in the history of humanity could not come to grips with a threat of its own making despite forty years of warnings about the consequences of business as usual.

I don't buy it. For one thing, implicit in this argument is the assumption that the wealth and power of modern civilization are based on one particular source of energy—fossil fuels. If true, that would be equally damning, because it would imply that luck, not human ingenuity, fed economic growth since fossil fuels came to dominate in the twentieth century.

On the contrary, there was nothing inevitable about fossil fuels. Most of the renewable sources being pursued today were under active development more than one hundred years ago, before fossil fuels crowded out all other energy sources except hydroelectric, and later nuclear. Our reliance on fossil fuels became self-reinforcing as cheap oil and coal in the early twentieth century created a massive installed base committed to further exploitation of these fuels. With economic power comes political power, and fossil fuels expanded their reach with the help of explicit and implicit subsidies. The explicit subsidies for oil have been whittled down in recent years, but they are still with us to the tune of about $85 billion for the United States and European Union.

It's the implicit subsidies, however, that show the true power of the fossil fuel lobby. These fuels are only cheap if we ignore the negative costs associated with their production and use. These costs range from air pollution; ecosystem damage; the poisoning of rivers, lakes, and oceans through oil spills; to debilitating and life-shortening respiratory ailments. And then there are the vast array of negative economic impacts associated with climate change. The International

Monetary Fund, hardly a hotbed of environmental activism, estimated that added together these explicit and implicit subsidies amounted to 6.3 percent of global GDP in 2015, or $4.7 trillion (and $649 billion in the United States). A good part of the implicit subsidy comes from the nearly 90 percent of the fossil fuel emissions that go into the atmosphere without paying any price at all, but that contribute not only to global warming but premature deaths from air pollution, impaired health, and other costs to society. A study involving a collaboration of researchers from Harvard and three British universities published in 2021 in *Environmental Research* estimated that pollution from fossil fuels accounts for 8.7 million deaths a year and has shortened the average life span by more than a year.

A group called Ecofys tried to show just the monetary costs imposed on the European Union in 2012 in a handy chart. They broke down the €250 billion by the source of the external cost:

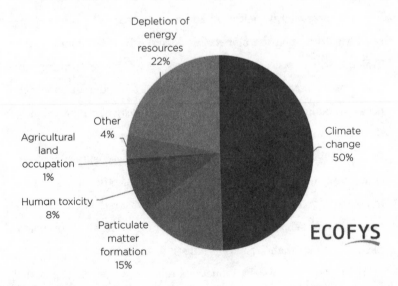

© Ecofys 2014 by order of: European Commission

Had modern economies used taxation to incorporate these costs into the price of fuels, society would have been shifting toward less-polluting renewables long before climate change became an issue.

Historically, society dealt with the "externalities" (as economists refer to these unaccounted costs) through legislation such as the Clean Air Act. As the scope of the threat of global warming came into focus in the 1980s and 1990s, it also became clear to those conversant with the problem that something was going to have to replace fossil fuels. The missed opportunities of earlier decades became irrelevant; the question was what could be done now or in the near future to meaningfully reduce greenhouse gas emissions.

Were alternative sources of energy ready to absorb the shift to renewables had society acted in the 1990s? The simple answer is yes. Early in the decade, there were a number of onetime opportunities to put the world on a different path. The dissolution of the Soviet Union in 1991 offered an extraordinary chance to reset what was then perhaps the most energy inefficient and greenhouse gas intensive economy ever to despoil the planet. Consider the East German car, the Trabant, which basically amounted to a device whose primary purpose was to create air pollution. For the cost of producing about ten times the pollution of other cars of its era, the lucky buyer got a vehicle that could accelerate from zero to sixty (its top speed) in twenty-one seconds, which meant that it would trail a good bicyclist until the very end. Simply modernizing Soviet bloc industry for efficiency offered a freebie that would produce both profits and a vast reduction in greenhouse gas emissions, which is why, after reunification, Germany could seem to both grow and reduce its emissions at the same time.

Estimates of emissions reductions related to the collapse of the Soviet zone run to 40 percent, and that does not even include such side effects as a decade-long reduction of meat consumption related to the economic collapse. A study published in *Nature* in 2019 estimated that the change in diet resulted in total emissions reduction of 7.6 billion tons of GHG in the ensuing decade.

Once European nations realized that they would at least have to appear to try to reduce GHG emissions, there was a scramble to get credit for this obvious way to game whatever treaty obligations might come in the future. These were real reductions, but they underscored another leitmotif of the global warming era—most nations would prefer to appear to achieve greenhouse gas reductions through accounting trickery than bite the bullet and actually do something.

The emissions gains achieved in modernizing the former Soviet states mostly had to do with energy efficiency, which could improve profits and reduce costs in a vast array of enterprises beyond the antiquated infrastructure of the Soviet bloc. Energy efficiency is perhaps the one genuine success story of the climate change era.

Amory Lovins, a physicist and founder of the Rocky Mountain Institute, has championed energy efficiency and renewables since the 1970s. He claims that gains in efficiency, particularly in reducing the energy intensity (the amount of energy used in a unit of production), have resulted in thirty times the GHG reductions of the adoption of renewables since 1995, and that these efficiency improvements are the reason that emissions have not increased as fast as economic growth in the United States. According to the American Council for an Energy-Efficient Economy (ACEEE), since 1980, U.S. GDP increased by 149 percent while electricity use only increased by 26 percent. Part of this had to do with structural changes in the economy (the shift from manufacturing to services), but a significant part came from increases in efficiency.

In 1990, Lovins coined the term "negawatt" to underscore how money could be made and pollution reduced by focusing on efficiency. These efficiency gains have continued through both Democratic and Republican administrations. This underscores the point that emissions reductions are possible when there isn't a powerful vested interest that loses when those gains are made. ACEEE asserts that energy efficiency in 2015 was the third-largest source of electrical power in the United States, ranking only behind coal and natural gas but ahead of nuclear power.

While most of the focus in the 1990s was on what the developed nations might do to reduce emissions, the real opportunity for dealing with climate change lay in the emerging nations and how they would power their industrialization and development. This was the true tragedy of the 1990s. The path not taken in the industrialization of the emerging nations turned out to be the world's best opportunity to reduce future emissions.

The concept, as expounded at countless conferences and international get-togethers, was to build out the energy infrastructure using a mixture of renewables rather than recapitulate the fossil-fuel-intensive path that the developed nations had followed during their industrialization. The phrase, repeated countless times in the 1990s, was to use "leapfrogging" technologies (it may well be that Thomas Johansson, the renewable energy expert, and colleagues introduced the phrase with regard to energy, as they used it in 1987 in an influential study on sustainable energy published by the World Resources Institute). There were many ways of doing this. Lovins argued that rather than building out expensive grids it made sense for the least developed nations to focus on distributed power generation using a mix of renewables appropriate to a region.

It didn't happen; no frogs leaped. The St. Louis Fed dates China's industrialization to 1988. Between 1990 and 2019, Chinese greenhouse gas emissions increased more than fourfold, rising from half the emissions of the United States to about twice our number (as noted, the United States has remained relatively stable thanks largely to massive improvements in energy efficiency). India's have more than tripled (it is now the third-biggest GHG-emitting nation in the world), as have Indonesia's.

Had China been incentivized to use a portfolio of renewables, that commitment would have provided economies of scale and incentives to accelerate technological progress in solar, wind, and other alternatives. The same would apply to other big emerging nations. Billions of tons of greenhouse gases that are now in the atmosphere might not have been emitted.

Instead, China and India heavily relied on fossil fuels for industrialization, and along with a changing climate got air pollution problems so severe that the mere act of going outside in their major cities now shortens your life. This was not inevitable. Leapfrogging technological progress has been accomplished in other areas. Much of Africa, for instance, has basically skipped the stage of land lines for telecommunications and gone straight to wireless.

Why this never happened with electrical power might be summed up in one word: coal. China had vast reserves in 1990, the fourth-largest in the world. If you didn't care about the costs to climate and health that came with burning coal, this domestic resource would be irresistible as a source of energy to power industrialization. And in 1990, China didn't care.

China produced and consumed about 1.2 billion tons of coal in 1990, toward the beginning of its massive push to become an

industrial power. Throughout the 1990s and 2000s it steadily increased its production and consumption of coal, with consumption
peaking in 2018 at about 4.6 billion tons, or nearly four times its consumption in 1990. To put this in perspective, the coal consumption of
the world's biggest economy, the United States, in 1990 was 896 million tons of coal at a time when the U.S. economy was between 5.5
(purchasing power parity) and 16 (on the basis of GDP) times the size
of China's. In 2018, with the U.S. economy still about 40 percent
larger than China's in terms of GDP, the United States burned 688
million tons of coal, or 15 percent of the amount burned by China.

The impact of China's choice of a development path has been
breathtaking. As COVID showed, China's air pollution can be cleaned
up in the blink of an eye. Once the carbon pollution migrates higher
into the atmosphere, however, it stays there for more than a human
life span.

Burning coal to release its energy results in the coal's carbon
binding with oxygen in the atmosphere. This produces CO_2 in a multiple of the weight of the coal burned. The rough figure one can use is
that burning 1 ton of coal produces about 2.9 tons of CO_2. China now
releases half the global total of more than 14 billion tons of CO_2 emissions from coal, with China's total CO_2 emissions from fossil fuels
coming in at about 10 billion tons annually.

Once China embarked on this path, it only required simple arithmetic to see what was coming in terms of greenhouse gas emissions.
India, another prime candidate for the adoption of leapfrogging
technologies, also chose coal, and its leapfrogging involved jumping
over Germany, Japan, and Russia to become the third-largest emitter
of greenhouse gases. Recall that the world was already seeing changes
in climate by the mid 1980s as a result of greenhouse gas overloading

in the atmosphere since the dawn of the Industrial Revolution. In other words, at a point when greenhouse gases already were having a meaningful impact on climate, China, India, and other emerging nations powered a massive pulse of industrialization with an energy source that was known to be the worst offender in terms of producing more greenhouse gases.

The impact of coal-related emissions comes from their accumulation over time. The oceans and the land seem to be able to absorb roughly 5 billion tons of CO_2 a year. Since 1990, China has averaged that amount of CO_2 emissions just from coal alone, with total annual GHG emissions soaring to 270 percent of earth's annual absorptive capacity by the late 2010s. To put this another way, China's choice of fuels for development has overburdened the globe's capacity to absorb CO_2 by increasing amounts every year since 1990, meaning that even if the entire rest of the world had reached net-zero CO_2 emissions thirty years ago, greenhouse gas concentrations would still have grown, and temperatures would have continued to rise. And, of course, China was not alone among emerging nations in choosing to power development with coal. In retrospect, given the consequences we are now dealing with, and given what was known at the time, the decision to use coal as the fuel for economic development by the big emerging economies has to rank as the most self-destructive choice in the history of modern civilization. Now a group of financial institutions, including Citi, HSBC, and BlackRock, are proposing to buy Asian coal-fired plants and retire them ahead of their normal schedules to reduce future emissions. This wouldn't be necessary if the plants hadn't been built in the first place.

So why didn't these emerging nations choose another path?

With regard to China, one surprising factor was the international and domestic shockwaves that followed after the government violently put down the pro-democracy protests in Tiananmen Square in 1989. Thomas Johansson was part of a Chinese council for international development cooperation that formed in the early 1990s. He described it as an attempt by the Chinese government to reengage with the international community after suffering widespread condemnation for its actions during the Tiananmen protests. The council consisted of about two dozen senior Chinese officials and an equal number of experts convened from around the world. Johansson was there to provide ideas about energy options. He remembers that in the years following the protests, the government became obsessed with maintaining stability, and the perceived key to stability was economic growth to provide food and jobs for their one billion people.

The pressure came not just from the trauma of Tiananmen, but from what ordinary Chinese saw going on around them. "By the 1990s, many Chinese had access to television," says Johansson, "and they saw how people lived in other parts of the world." The officials were aware of environmental issues, he says, but stability came first. "They wanted to learn from the West about renewables, and they did learn," says Johansson, "but it took them a while to catch up."

In the meantime, China turned to coal as the fuel for the double-digit economic growth that was their goal. The path to coal was also greased by corruption. A large coal-fired power plant offers a more tempting opportunity for a "commission" than smaller scale and scattered renewables projects, says Johansson, particularly as there were fossil fuel interests ready to open their wallets.

Another issue, he says, was that the West was completely tone-deaf in the way that it tried to convince other countries to adopt renewables. While the Chinese were worried about pushing economic growth immediately, they were being told by the West that they should be limiting fossil fuels to prevent a problem in the future. Instead, Johansson argues, renewables and efficiency should have been pushed as a way to spur economic growth at the local level. "It would have converted a long-term discussion into a near-term local benefits issue," he says. "If you do energy right, you solve a host of problems—local development, energy security, health benefits—and get climate mitigation as a side benefit."

It didn't help matters that the developed nations were basically telling the Chinese and Indians to "do as we say and not as we do." The Chinese looked at the West and saw that renewables were not a top priority, even though up to that point it was the developed countries that had created the problem of overloading the atmosphere with greenhouse gases. In the early 1990s, emissions per capita in the United States were nine times those in China and twenty-seven times those of the average Indian. And yet we were telling them not to use the fuels that we used to power our economic growth and also telling them to spend their own money developing the alternatives. Worse, they knew that there wasn't even a consensus in the United States that global warming was a threat to be addressed. Given all this, opting for renewables was a deal they could refuse, even if it meant that down the road we all would suffer.

China, India, and other emerging nations could have industrialized using far less coal. In 1990, China emitted more greenhouse gases to produce one dollar of GDP than any country in the world. The same was true in 2014. In 1990, the United States used little more than one-third the emissions of greenhouse gases to produce one dollar of GDP than China did. By 2014, China had narrowed that gap but still created nearly twice the emissions of the United States to produce a dollar of GDP. That same year, France, with its heavy reliance on nuclear power, emitted nearly 80 percent fewer greenhouse gases to produce a dollar of GDP. Coal was not and is not a necessary evil for economic development.

Still, if China, India, and other nations had adopted an alternative path to industrialization in the 1990s, it would have required a united front of the developed world to create the mix of incentives and disincentives to encourage these countries to pursue an energy source other than the massive reserves of coal located within their borders. As noted, something close to the opposite was the case.

This is where public opinion comes into play.

In 1993, the year rapid climate change became an ominous possibility, the newly inaugurated Clinton administration got a brutal baptism in how hard it would be to do something about greenhouse gas emissions. President Clinton's first budget contained a proposed BTU tax that would tax fossil fuels while exempting alternatives such as wind and solar. It was billed as a revenue-raising measure that would encourage energy conservation, reduce pollution, and reduce dependence on foreign oil. Inside the administration, it was also viewed by Vice President Al Gore and others as a vital step in dealing with the recently emerged threat of climate change.

It never had a chance. According to Christopher Leonard's *Kochland: The Secret History of Koch Industries and Corporate Power in America*, the oil tycoons had begun preparing to spread confusion and denial as early as 1991, when they sponsored a Cato Institute meeting that brought together climate change deniers and like-minded business interests to plan how to derail action on global warming. When the BTU proposals arose, the fossil fuel lobby was prepared.

Opposition arose from all quarters. The National Association of Manufacturers, the American Petroleum Institute, the U.S. Chamber of Commerce, and farmer associations banded together and spent $2 million on ads on how the tax would hurt the economy. The Koch brothers poured millions more into another of their astroturf organizations, Citizens for a Sound Economy, to pressure key politicians such as David Boren, the senator from Oklahoma (an energy state) who chaired the Senate Finance Committee.

Most of the backlash focused on costs rather than anything related to climate, but the message was loud and blunt: threatening fossil fuel interests puts one's political future at risk. Anthony Leiserowitz says the backlash almost cost Clinton his ability to govern. A number of congressional representatives lost their seats in the subsequent midterms in 1994, and many were said to have been "BTUed." Along with Clinton's ban on assault weapons, which galvanized the National Rifle Association, the BTU tax was one of the factors that swept Newt Gingrich into power, basically neutering the possibility of any grand initiatives Clinton might have pursued in other areas like health care.

If you were a political creature in the mid-1990s, the lesson of the BTU debacle was that there was a new third rail of politics. Whatever momentum climate change might have had with the incoming Clinton administration dissipated with that defeat. Worse, the reaction poisoned political opinion on using taxes to address climate change, thus nullifying perhaps the most efficient way to address the problem.

The BTU tax debacle showed the power of the fossil fuel lobby, and now that the beast was awakened to the threat of climate change action, it quickly adapted the playbook perfected by lobbyists during the political battles over tobacco and the ozone hole. The climate change denial movement was born. Just as a hurricane draws power from the energy stored in warm ocean waters, the climate denial movement found wellsprings of energy in the individualistic, anti-government, anti-elitist streak in the American public.

The Republican Party had begun its move toward libertarianism during the Reagan years, and now GOP politicians were being asked to support global action to face an evolving threat. Tim Wirth, a Democrat who chaired the Senate hearings on global warming in 1988, remarked to me in the 1990s that climate change was one issue on which Republicans "lose all reason," because they saw it as a Trojan horse for world government.

In a recent conversation, Anthony Leiserowitz expanded on that thought. He described climate change as "the policy problem from Hell.... You could not design a problem less suited to our psychology and our politics." For one thing, as the GOPers feared, "it was the mother of all collective-action problems—90 percent of the solutions involve structural change that we can only do collectively," meaning with government involvement. Dealing with the problem also required treaties and international cooperation, and with that came issues of agreements impinging on U.S. sovereignty in a host of activities.

Leiserowitz says that it is no coincidence that the denier movement is most evident in four English-speaking countries: the United States, Canada, Australia, and England. Three of those countries are former colonies with recently conquered frontiers. With that history came strong subcurrents of individualism. All four are major fossil fuel producers. "If your worldview is focused on individual freedom and is characterized by an antiregulation, anti-taxation bias," he said, then addressing the problem is your worst nightmare. "On the one hand they don't believe the threat, but rather see *action* to avert global warming as the real threat." Using the analogy of explosives, he said that "uniquely, the U.S., Canada, and Australia have the nitro-abundant fossil fuels and the glycerin-receptive politicians" as the ingredients of an explosive situation. He argues that there was no better distillation of those two strands than the network funded by the Kochs.

Little wonder, then, that the emerging nations saw hypocrisy and mixed messages coming from the West, particularly the United States, on climate change. Moreover, from the Western point of view, the stumbling blocks to collective action were not trivial. The biggest was summed up by the phrase "free rider." The big emerging nations said, in effect, why should we limit our use of cheap fossil fuels when we didn't create the problem, you did, and our per capita emissions are a tiny fraction of yours? The developed nations' retort was that the world could stop climate change only if all countries limited emissions, and the big developed economies should not be forced to put their exporters at a structural cost disadvantage to those in emerging economies.

Lovins, Johansson, and others would say that this was a false choice—that the emerging economies could have developed along other energy paths (with Western help). That message, however, never got traction in the 1990s. Johansson believes that one reason is that the fossil fuel industry successfully framed the problem as dealing with a far-off, unproven threat, rather than an alternative, economically viable way to develop.

Negotiators didn't solve the free-rider problem in the 1990s and still haven't solved it. One solution was and is that any treaty needed to be compulsory, but of course that raised issues of sovereignty. To get a sense of the degree to which the free-rider issue gained traction in U.S. domestic politics, consider what happened to Al Gore before he went to Kyoto to negotiate binding emissions on behalf of the United States. He was hobbled by the Byrd-Hagel Resolution (passed 95 to 0), which held that the United States could not be signatory to any mandatory reduction of greenhouse gas emissions unless the agreement held developing nations to similar reductions in the same period.

I looked at the impasse on climate negotiations and proposed my own solution, which I felt dealt with the free-rider problem and brought the developed and emerging economies into the same boat. I published it as an op-ed in *The New York Times*. In a nutshell, I suggested coming up with an overall goal for emission reductions, dividing the world into three giant regions that included both developed and emerging nations, and then letting those regions achieve an overall reduction anyway they wanted.

Two weeks after this op-ed came out, I participated in President Clinton's White House Conference on Global Climate Change. Such is the White House's convening power that the participants included an elite group of economists, academics, scientists, and industry leaders, really all the people you would want at a conference to mobilize action on climate change. Nothing was mobilized. Why not was summed up in an almost bemused comment by President Clinton. Echoing FDR, who famously remarked to a group of activists, "You've convinced me, now go out and make me do it," Clinton remarked that he wanted to take action on climate change, but that he could not get ahead of the public on the issue, basically telling the participants that there was no public pressure for him to do anything and he couldn't do anything without it.

One of the goals of the White House conference was to educate the public about the threat so that they would, in fact, make Clinton act on climate change. It didn't work. Polling out of Stanford University showed that while Democratic concern about climate change remained roughly unchanged before and after the education outreach, GOP concern actually dropped from 67 percent to 55 percent. Much of that drop was attributed to increased partisanship in relation to the Clinton impeachment. It speaks volumes about the fragility of momentum on climate change action that it could be derailed by a president lying about sex.

So it was that the 1990s ended with the scientific clock still trailing reality but rapidly catching up, while the clock of public opinion had barely started ticking. Most in the developed world thought the problem was way off in the future, while those in emerging nations were largely unaware that climate change was even an issue.

Ten

BUSINESS AND FINANCE IN THE 1990s: Mobilizing, but Against Climate Action

The business and finance community got what it wanted on climate change in the 1990s—no action whatsoever. This was the result of a many-pronged effort, ranging from influencing the structure of the initiatives to combat climate change, torquing the agendas of international conferences, putting pressure on politicians, and fueling the nascent efforts of what was to become the climate change denier movement.

These "successes" began early in the decade when energy was dropped as a separate section in the Rio Earth Summit of 1992. In 1990, a group had been formed at the recommendation of the UN General Assembly to prepare a report on renewables as part of a proposed chapter on energy in Agenda 21, the road forward that was to be agreed upon at the Rio Conference. Thomas Johansson chaired the group, and the report was delivered to the relevant preparatory committees. Then nothing happened. The commission took note of the report, but mysteriously—it was never explained to Johansson—energy disappeared from the agenda. The conveners decided that they didn't need a separate section on energy solutions at a conference whose primary focus was climate change and development. This was akin to convening a conference on pandemics and not mentioning COVID-19. To this day, Johansson is convinced that the fossil fuel industry lay behind this omission, abetted by the oil-producing nations and a vast network of backroom influences. Johansson notes that it took another twenty years for a Rio Conference successor to explicitly include renewable energy as a sustainable development goal.

Other factors also slowed the movement toward renewables. In the United States, major capital projects such as a power plant typically have a life span of roughly thirty years. In the early 1990s, a large number of coal-fired plants were coming to the end of their useful life. This meant that whatever technology replaced them would be with us for another thirty years. In effect, the utilities would be making a choice about the sources of electricity for the next generation. They didn't choose renewables. Any impetus to shift to renewables was short-circuited because natural gas was plentiful and thus prices were quite low in the early part of the decade. Moreover, combined-cycle plants (power plants that use both furnace heat and waste heat to generate electricity) were getting much more efficient. This precipitated a massive shift to gas-fired power plants, committing many utilities to fossil fuels until the third decade of the new millennium.

This was just bad luck. Other setbacks were planned hits. The dense, hedged wording of the Executive Summaries produced by the IPCC, the beatdown Clinton suffered on the BTU tax, and the snubbing of energy at the Rio Conference all bore the fingerprints of industry pushback. At the beginning of the decade, those who wanted action on climate faced a virtual united front. Pillars of the business establishment such as the U.S. Chamber of Commerce and the National Association of Manufacturers joined with traditional adversaries such as the unions, including the United Auto Workers, the AFL-CIO, and the United Mine Workers, to oppose the UN-sponsored treaty process on climate change.

Economics could have helped but didn't. Environmental economics enjoyed a renaissance during the 1990s, as many creative minds tried to think of ways to price environmental services and the cost of pollution and otherwise bring economics into harmony with ecology. It was enormously important work.

Few governments listened. Business listened, or at least their PR departments did. British Petroleum, the oil major, announced that BP stood for "Beyond Petroleum," and, for a few years, the company became the darling of environmentalists. Even as it was advertising its green credentials, however, the company was supporting fossil fuel industry efforts to weaken regulation of offshore drilling, successful efforts that led to the Deepwater Horizon oil spill of 2010, the largest in U.S. history. Efforts to block regulatory strengthening of the Minerals Management Service, the agency responsible for regulating offshore platforms, began when reforms were announced in 1991, and the delaying tactics were so successful that the reforms weren't finalized until nearly twenty years later when the spill took place.

Economists did try to tackle the future costs of climate change. The hope was that by bringing the alleged rigor of economics to calculating the future costs of various scenarios for global warming, policymakers could make informed decisions about costs and benefits of future actions. Based on the models developed by mainstream economists in the 1990s, the informed decision would have been to do nothing, a message that was taken up with a vengeance by the business community.

Richard Tol, a British environmental economist, published a study in 2018 entitled *The Economic Impacts of Climate Change*. In it, he looked at twenty-seven estimates of future costs of various degrees of warming that had been published since the 1980s. Each estimate gave a number for the average reduction in a person's income for a given temperature change by 2100. Perhaps the most prolific and influential modeler was William Nordhaus of Yale, a respected economist whose efforts to price the cost of climate change dated back to the 1970s. For an anticipated 3 degrees Celsius of warming, his best estimate, published in 1991, was a 1 percent loss of income. In 1994, he published two estimates with most likely being a 1.3 percent loss in one and a 3.6 percent loss in the other. In 1996, he and a colleague published an estimate of a 1.4 percent loss for a 2.5 degree Celsius increase in global temperatures.

One reason for such modest estimates was that Nordhaus assumed that because most of the U.S. economy happened indoors—87 percent was the figure he arrived at in 1994—it was immune to the impacts of climate change. It's mind-boggling that a serious scientist would have made that assumption. The huge petroleum and chemical facilities in Houston and New Orleans probably did not feel immune after Hurricanes Katrina and Harvey, and the myriad indoor businesses of the U.S. West might dispute their immunity to the impact of the wildfires of 2020.

If economists ignored second-order impacts, climate scientists did not. Nordhaus requested estimates for potential economic damage from a number of prominent climate scientists. Their estimates came in twenty to thirty times higher than those of the mainstream economists.

Nordhaus's estimates for the hit to the U.S. economy were even more modest. In his 1993 paper "Rolling the 'DICE': An Optimal Transition Path for Controlling Greenhouse Gases," he wrote, "A growing body of evidence has pointed to the likelihood that greenhouse warming will have only modest economic impacts in industrial countries, while progress to cut GHG emissions will impose substantial costs." How modest? Nordhaus estimated that a 3 degrees Celsius warming would cost the U.S. economy a minuscule one-quarter percent of national income. He admitted the possibility of unmeasured or unquantifiable variables he might be missing, but in his view, they might only bring the cost up to about 1 percent of national income. To put this in perspective, last year Moody's Analytics estimated that the global economic toll of 2 degrees Celsius warming was $69 trillion. A recent study undertaken by Oxfam and Swiss Re estimated that the costs to the global economy of a 2.6 degrees Celsius rise by 2050 would be 13.9 percent of GDP each year (relative to a world without warming), an estimate that is three times the damage inflicted by COVID at the height of the epidemic. The damage from 3 degrees warming is likely incalculable, as ecosystems would fail and civil order disintegrate in many places.

The do-nothing crowd took Nordhaus's estimates and ran with them. In 1997, for instance, the late William Niskanen, then chairman of the ultraconservative Cato Institute, seized on Nordhaus's figures to argue before Congress that it was premature to take action on climate change because "the costs of doing nothing appear to be quite small."

These estimates from the 1990s become even more absurd when put in the context of the rapidly advancing science. Four years after climate scientist James Hansen told Congress that global warming was already happening, Nordhaus suggested that the "thermal inertia of the oceans" meant climate change would have a "lag of several decades behind [greenhouse gas] concentrations." When I asked Thomas Johansson how Nordhaus could talk about a lag of several decades before the impact of global warming became evident when record hot years were already accumulating, he replied, dryly, "Only an economist can do that." Partly for his efforts in the economic modeling of climate change, in 2018 Nordhaus was awarded the Sveriges Riksbank Prize in Economic Sciences in Memory of Alfred Nobel, commonly referred to as the Nobel Prize in Economics.

There was one very large sector of the economy that did not buy these benign estimates of future risk: the property and casualty insurance sector, particularly the reinsurers who insure against catastrophic losses. Because they live or die by accurately pricing risk, the expectation was that these companies would be hypervigilant in surveying for risks just beyond the horizon—such as climate change— and that they would be active in lobbying Congress for action, just as they had been for lighting standards, seat belts, and construction standards in storm zones.

It didn't happen, and therein lies a mystery. A $2 trillion industry, the insurers certainly had the heft to get a hearing. And, in fact, the reinsurance industry took an early and deep look at the risks associated with climate change. Giants such as Swiss Re and Munich Re sponsored conferences and reports on how global warming might inflict economic damage. I was sufficiently impressed with the apparent commitment of the insurance industry to lead climate action that

I wrote an article for *Time* in 1994 about the positive role that insurers might play in focusing policymakers on the financial risks of climate change.

It wasn't that reinsurers weren't aware of the risk. In that article I interviewed Frank Nutter, then and now president of the Reinsurance Association of America. "It is clear," he remarked, "that global warming could bankrupt the industry."

Insurers also had the perfect set of tools to force people to confront the issue. If the reinsurers began to raise premiums or cancel policies for homes and businesses at heightened risk for sea level rise, more frequent and intense storms, and wildfires, it would bring home the message that climate change was a pocketbook issue. Insurers did pull out of some markets in the 1990s, but underwriting continued apace in many other at-risk areas. Nor did insurers put pressure on Congress to act as they had in the case of seat belts. Worse, the big reinsurers continued to underwrite policies for coal-fired plants (without such insurance, most plants could not get financing), which meant that even as they were estimating the negative economic impacts of climate change, they were enabling further emissions from the worst greenhouse gas offenders.

It would be more than twenty years before reinsurers began in earnest to take the actions that I and others expected in the 1990s. The delay resulted from several factors, some of which might have been anticipated, others not. The case study of how the insurance industry has responded to climate change is disheartening and also a cautionary tale because it shows how perverse incentives can thwart action even in an industry exquisitely tuned to risk.

I made three mistakes in 1994. The first was forgetting that insurers have to sell insurance. I underestimated the degree to which incentives at the retail end of the property and casualty insurance business differed from those at the other end—the part that dealt with catastrophic risk. The second was to underestimate the industry's genius for coming up with ways to spread, shift, and otherwise defang risks. The third was to underestimate the degree to which embedded perverse incentives up and down the chain made it imperative for all to continue to compete for business even as the risks of climate change became more obvious.

By the early 1990s, the insurance industry "knew" some basic, scary aspects of climate change. They knew that it was likely that there would be an increase in frequency and intensity of extreme weather events and that this reduced the utility of using past patterns and data for predicting future losses. The industry also knew that they could not use straight-line projections to predict future losses; that the interaction of an increased number of extreme events with social, economic, and political factors made such predictions nonlinear in the sense that the secondary reactions to events would carry with them unmeasurable and unpredictable costs as well.

Given that, in the early 1990s, it was natural to expect that property and casualty lobbyists would try to influence politicians to take actions that might head off or mitigate the threat, as the industry did with seat belts, smoking, and electrical standards. That did not happen, at least not with anywhere near the vigor that the industry applied to the other problems.

It would also be natural to expect that insurers would begin to pull back or reprice policies in areas most at risk for extreme weather. That happened a bit, but for complicated reasons, not nearly to the degree one might expect.

In sum, where the industry in the past had been proactive with risks that might affect its bottom line, with regard to climate change it has been reactive. The reason has to do with the structure of the industry. Thus, an industry that should have acted as an early-warning signal of climate danger, and also, through its pricing, might have forced action by pricing the threat, failed to deliver either the early warning or the pricing signals that might have made a difference in the 1990s.

Let's begin with a brief tour into the plumbing of the insurance industry. When a homeowner buys an insurance policy, the agent who sells the policy gets a commission. The company that underwrites the policy will backstop the losses up to a certain point, and other companies will join in to pick up the next tranche of losses. Most of the last tranche of losses, those that come from true catastrophes, will be picked up by the reinsurance industry.

The agent who sells the policy doesn't really care about catastrophic risks. So long as the company the agent works for is willing to back the policy, the price presumably accounts for the risk, and until catastrophe actually strikes, the agent's loss ratio (the proportion of policies written that involve claims) will be good and he or she can participate in profit sharing.

Chris Walker, who now promotes environmentally sustainable investment, used to be a managing director in Swiss Re's sustainable business unit, where he was in charge of Greenhouse Gas Risk Solutions. Swiss Re is a giant in the reinsurance industry, and from that perch Walker had a synoptic view of the force field of competing incentives that swirled around the pricing of climate change risk in underwriting insurance policies. He agrees that the incentives at the retail level were tightly focused on selling policies. This meant keeping prices competitive with other insurers. Because the rewards of writing policies were immediate, and the costs of realized risks lagged, this meant risk was habitually underpriced.

In 2008, the world got an expensive lesson in what happens when perverse incentives lead the sales end of an industry to underprice risk. That financial meltdown provides a foretaste of what might happen in the insurance industry in coming years.

As noted in the introduction, one of the triggers of the 2008 crisis was the earlier parabolic rise of home prices as the explosive growth of a new type of bond incentivized lenders to write mortgages regardless of whether a household had the wherewithal to keep up payments. The bankers got fees for writing mortgages but didn't worry about repayment, because they quickly sold the mortgages to other bankers who would use the mortgages as the collateral for billion-dollar securities. Those buying the mortgages had a false sense of security because data going back to World War II showed that home prices rose only on an annual basis.

Insurance brokers are similarly incentivized to continue to write policies until a catastrophic event happens. Chris Walker says that typically there is a year lag between the event and when losses appear. Consequently, even though insurance risk analysts knew that climate change increased the odds for western fires, until serious outbreaks of such fires occurred in the 2010s, the incentives for the agents at the retail level were to continue to write policies in the fire zones.

Competition for business also explains a reluctance for reinsurance companies to tighten standards. Ordinarily, when a new risk surfaces, the reinsurers will first ask about the insured party's possible exposure to the risk and what they are doing about it. If the risk is broad enough and expected to continue—e.g., sea level rise and increased storm frequency and intensity—the reinsurers will pull back from underwriting specific policies or add exclusions for certain kinds of risks.

At all levels, from retail to reinsurance, insurers have an ace in the hole. Most policies have to be renewed yearly, which gives an insurer some comfort. Instead of having to consider a vastly increased likelihood that a major hurricane or wildfire will hit a specific area at some point in the future, they can limit their worry to whether a catastrophe will hit a certain property in the next year. A house that has a 100 percent chance of being flooded in the next hundred years has only a 1 percent chance of being inundated in the next year. And if because of climate change such floods become twice as likely, that risk doubles—but it is still only 2 percent.

Still, reinsurers are not always eager to absorb a doubling of risk. This is where industry ingenuity came into play. Ironically, it was a major hurricane that led to an innovation that enabled the insurance industry to continue underwriting policies in areas at risk for climate change.

The hurricane was Hurricane Andrew, which hit just south of Miami as a category 5 hurricane in August 1992. During its short, thirteen-day life, the monster storm wrecked 125,000 homes in Dade County alone. The destruction led to the insolvency of eleven insurance companies, and it could have been worse, since it was a relatively compact storm and hit shore a few miles south of a major city. Estimates of the damage of a direct hit to Miami today range from about $50 billion to more than double that number.

After that debacle, reinsurers began balking at fully backstopping risk, and sixty-three insurers either left the state or curtailed new business. A mathematician named Eberhard Müller at the German reinsurer Hannover Re began thinking about possible solutions and came up with the idea of a security, which became known as CAT bonds, that allowed outside investors to get paid generous interest rates to take part of the risk. It was a brilliant idea from a number of perspectives. It allowed reinsurers to offload some of their risk so that they could continue to underwrite lucrative policies in at-risk areas. It also gave reinsurers access to the vastly larger financial resources of the world's markets. For investors—usually hedge funds and institutional investors—it offered tempting returns that were uncorrelated to other markets (in the sense that the risks of these bonds had nothing to do with market movements or the economy), allowing them to diversify their portfolios.

An obvious question was, why would any investor assume a risk that very smart reinsurers wanted to unload? The answer has to do with the aforementioned limited time exposure of the risk. Even if climate change dramatically raises the probabilities of catastrophic events, the odds of a particular catastrophic event at a particular place within the time frame of a few years don't really rise that much. A CAT bond might insure the risk of a category 5 hurricane hitting a specific area like Miami in the next two or three years. Another reason is the recent global financial context of ultralow interest rates. In a world awash with negative interest rates, the bonds offer institutions and hedge funds fat returns.

In terms of climate change, the invention of CAT bonds allowed insurers to put off facing the issue. As Chris Walker noted, "Just as was the case with mortgage securitizations in 2007, your underwriting criteria don't have to be that strict if you're going to be offloading that risk."

Walker's job was to incorporate the increasingly scary findings on climate change into underwriting. He says he never got traction except in some cases involving climate liability for directors and officers (called D&O insurance). For example, he noted that if a given company was 1 percent of global emissions and knew for thirty years that climate change was a real threat, that company might be found to have 1 percent of the liability. Walker said that they did manage to get some climate liability exclusions written into some D&O policies. He fully expects that some suits will be forthcoming. "Look what happened with tobacco," he says. "It took one case to break through and then the floodgates opened."

As it turned out, reinsurers had several ways of kicking the ball down the road rather than integrating climate risk into their insurance pricing. Apart from CAT bonds and limiting policies to one year, they could also pull out of areas deemed too risky. That's what happened in Florida after Hurricane Andrew. Even as the risks grew for coastal areas, those same areas enjoyed a building boom as affluent and aging Americans sought out the sun coasts, the closer to the water the better. Insurers sought to raise rates, but state regulators wouldn't let them. A number of big insurers responded with a collective shrug and said sayonara.

In the following decade, the Republican-led government of Florida chose to protect its coastal citizens from having to pay the true price of living in the crosshairs of wind risk by hurricanes by socializing that risk. This in turn had the effect of camouflaging the risk and encouraging people to move into harm's way. Current statistics are that 5 million people have moved to the Florida coast *since* Andrew.

Hurricane wind risk is but one threat global warming carries with it. Another is flood risk from sea level rise and storm surges. Here too insurers routinely exclude flood and storm surge damage from policies. Again, if that risk was properly priced, buyers might think twice about purchasing homes in harm's way and thereby become aware that global warming could cost them money. But there's profit to be made by burying that thought, and so that risk has been socialized at the federal level through FEMA's National Flood Insurance Program. Here too the government underprices risk. In fact, flood insurance policies not only fail to reflect risks priced on recent history, they are based on maps that don't reflect the realities of sea level rise (in 2019, the Trump administration tried to stop FEMA from updating their maps). In this case, it's the entire nation that subsidizes those who choose to live in harm's way.

Thus, a combination of perverse incentives, the efforts of reinsurers and states, and the unintended consequences of federal programs all served to dampen a price signal that might have caused people to take notice of the threat of global warming. Had insurers done in the 1990s what many thought they would do, it would have mooted the Panglossian economic forecasts of Nordhaus. It would have let the public know that global warming was not just some hypothesis about a far-off danger but a pocketbook issue. The reality that this threat carried costs in the present would have undercut the disinformation campaign of the fossil fuel industry.

Underwriting might sound like an arcane corner of the financial world, but the consequences of the failure of the insurance industry to price the risk of climate change were global. Millions of people facing rising insurance costs in coastal areas, or areas at risk for wildfires, would have made adjustments in their choices about where to live and likely also made climate change a political issue. The insurers might have opted out of underwriting coal-fired power plants much earlier than they did. Investors might have put more money into alternative energy and governments created more incentives for the shift to renewables. Perhaps there would have been more of a sense of urgency at the Kyoto talks later in the decade.

There was one very positive innovation in business and finance in the 1990s with regard to climate change, although it only had real impact starting in the 2000s. This was the adoption of so-called feed-in tariffs, first by Germany, then other European countries, and then, starting in the 2000s, by China, India, and other big emerging economies. As of this writing, the United States is the only major economy not to have adopted feed-in tariffs as a way of spurring investment in renewables.

Here's how they work. The tariffs, as they are presently configured, provide a guaranteed return to investors over a period, usually twenty years. This structure takes much of the risk out of financing wind, solar, and other renewable energy projects. Unfortunately, it took a decade of tinkering to find a structure (now the cost of generation plus a reasonable rate of return) that mitigates the uncertainties and unpredictability of fluctuating electricity costs. The new model has proven wildly successful. These successes underscore the crucial role that finance has played in the global warming era.

But in the 1990s, moneyed interests were a distinct negative, although they don't bear all the blame for the world's inaction. The IPCC, the natural reticence of scientists to go beyond the data, the emergence of a well-funded disinformation campaign, lack of charismatic political leadership—all deflated any sense of urgency that might have been felt about the issue.

It could have been otherwise; the world was not fated to fail to address the threat of global warming. Instead of a lost decade in terms of taking action, the 1990s could have been the decade during which the developed world began the shift that is underway now, and it could have been the decade during which emerging nations powered their development with leapfrogging technologies. Had this happened, we might have entered the new millennium with true momentum behind a shift away from fossil fuels.

Instead, the new millennium began with the reality of climate change already upon us, with the scientific community fully aware of the imminence and degree of the problem but with the public still barely taking notice, and with the financial community still living in a world where the biggest perceived issue was the imagined threat to profits that might come from action to head off global warming.

The failure of the United States and the global community to address the problem has larger implications than putting us on a ruinous path toward climate change—though that one consequence is plenty bad in itself. Climate change posed a test: Could the world's biggest economies change course to avert a global catastrophe? They couldn't.

Those same countries passed this test in the 1980s when nations came together (mostly) to defuse the threat of CFCs to the ozone layer. In that case, however, the major special interest, DuPont, had an incentive to support global action. In the case of climate change, by contrast, a wide spectrum of special interests were driven by financial incentives to continue business as usual. If the dominant moneyed interests see global action as endangering short-term profits, they will thwart collective action, even if it increases the likelihood of long-term disaster. There couldn't be a better example of an economy designed to drive off a cliff, and we've seen it borne out time and again since the 1990s—in the tech crash of 2000, the Great Financial Crisis of 2008, and, just recently, in the U.S. reaction to COVID-19.

PART FOUR

The New Millennium

Eleven

REALITY IN
THE OUGHTS

G uided by geophysics and geochemistry rather than politics and special interests, the changing climate continued to demand notice as the new millennium began. Seven years after the collapse of the Larsen A Ice Shelf, Antarctica's Larsen B Ice Shelf disintegrated in 2002, removing an expanse of ice the size of Rhode Island from the Antarctic Peninsula. The ice shelf was already floating, and so its collapse did not raise sea levels. On the other hand, the shelf had held back the advance of the many land-based glaciers behind it, and once that cork was removed, the movement of those glaciers, whose ice does raise sea level when it reaches the ocean, accelerated by a factor of eight, giving scientists a firsthand view of one way in which changes in the Antarctic might rapidly raise sea levels.

In fact, iceberg calving and meltwater runoff from the great ice sheets of Antarctica and Greenland were already having a measurable effect on sea level rise (although it would be more than a decade before the IPCC would acknowledge their contribution to sea level rise in its Summary for Policymakers). Recall that in the first IPCC report in 1990, some scientists believed that the Antarctic ice sheets might expand over the next hundred years, which would have had the effect of lowering sea level (the logic behind this expectation was that rising temperatures in the cold regions might have the impact of increasing snowfall, which would ultimately thicken glaciers and ice sheets). Instead, by the new millennium, the shrinking ice sheets were becoming a significant factor in the acceleration of sea level rise.

Sea level rise is the most unambiguous signal of a changing climate simply because only a very few factors can influence it on a global scale. Changes in temperature in the oceans cause the volume of water to expand during warming periods and contract during cooling. During cool periods land ice increases, sequestering water that would otherwise flow to the oceans. During warm periods glaciers and ice sheets release that sequestered water as meltwater or through the increased calving of icebergs. Global sea level nets out all the various inputs, and it can be measured by satellite. If it is rising on a decade-by-decade scale, then the climate is warming.

Moreover, because most coastlines slope gradually, a small rise in sea level can have a dramatic impact on flooding and beach erosion. Florida is the flattest state in the nation, and estimates are that a further 1-foot rise in sea level could move the shoreline between 2,000 and 10,000 feet inland. Sea level has already risen about 8 inches in the past thirty years, and given the accelerating rate of this increase, the next foot of rise might happen in the next thirty years.

By the middle of the first decade of the 2000s, sea level was rising at roughly twice the rate of the decade before, and the rate of rise continued to accelerate. While previous sea level rise had been driven by thermal expansion as the oceans absorbed heat, during the oughts, meltwater from glaciers and ice sheets began to surpass thermal expansion as a contributor, though that only became clear to those studying sea level rise several years later. Moreover, given the inertia of the massive ice sheets, once they began to meaningfully contribute to sea level rise, inevitably that contribution would continue to increase for the foreseeable future regardless of any action humans might take to reduce emissions.

While glaciologists and oceanographers were struggling to find ways to monitor and analyze the contribution of ice sheet melting and iceberg calving to sea level rise, the effects of rising seas were already being felt in coastal communities. A study of sea level rise and flooding led by the University of Miami found that the frequency of floods caused by tides rather than rain in Miami Beach increased by 400 percent after 2006.

Perhaps the most dramatic indication that something was changing was the advent of floods that bubbled up from storm drains on days with clear skies. Dubbed "blue sky" and "sunny day" floods, these started happening in coastal towns in the Southeast during what were called "king tides." In some cases, these were exacerbated by land subsidence as communities drained groundwater, but in all cases sea level rise was a factor.

Somewhat befuddling, however, is the fact that even after blue sky flooding became common, building in coastal Florida and other eastern seacoast areas continued to boom. After the housing bust of 2007–2008, Miami Beach and other coastal cities rebounded strongly, with pricey new waterside developments going up even as their affluent owners had to wade through occasional king tides to get to the lobby. Some of this might be explained by the complicated factors that made Florida a destination for flight capital from Russia, Asia, and South America. Still, one wonders what was going through a buyer's mind if they still purchased an apartment after reading about king tides. Did such buyers think there will be fewer such events in the future?

Sea level rise provided a subtle, ominous reminder that climate was changing during the oughts, even if most of those in harm's way blithely ignored the warning. Nature also provided other, more dramatic warnings that change was afoot. In 2003, Europe suffered the worst heat wave in nearly five hundred years. Some thirty-five thousand died as a direct result of the heat wave, and, subsequently, an additional thirty-five thousand deaths have been attributed to the sweltering temperatures. This event was the first of several major heat waves to hit Europe since 2000. Record-setting heat waves also hit parts of North America, Australia, and Asia, and the decade ended with an intense heat wave that impacted the entire Northern Hemisphere.

And then there were the storms. On August 29, 2005, Hurricane Katrina hit New Orleans after barreling across the southernmost part of Florida. It had lost some of its punch, top wind speed having dropped from 174 miles per hour to 125 miles per hour, but a concatenation of other factors gave the United States and the world a lesson in the damage that can occur when a major hurricane hits a major city.

New Orleans is a gift of civil engineering, as much of the city lies below sea level. The Mississippi River passes through as an elevated highway (as John McPhee memorably described it), with the river and tidal areas walled off from the city by hundreds of miles of levees. Before Katrina, many worried that the Mississippi River posed the biggest threat to the city. Instead, it turned out that the inundation came from the south, in the form of a 19-foot storm surge. Throughout the city and surrounding areas, the levees proved inadequate. Some were overtopped, one was rammed by a loose barge, others fell victim to scouring and were undermined by the erosive power of torrents of water, and others failed because engineers had overestimated the density and composition of the soils at their base. Twenty-eight failed in the first day, a number that soon grew to fifty.

The ensuing damage was biblical. More than 70 percent of the city's houses were damaged, with 56 percent suffering major damage. Eighty percent of the city was underwater during the worst of the flooding, and St. Bernard Parish was entirely submerged. With no place to live, half the population left. Before the hurricane New Orleans had a population of 452,000. In 2019, fourteen years later, the population was still lower by 160,000, 35 percent less than it was before the storm.

Economically, Katrina was the costliest natural disaster in U.S. history. Total damage resulting from the storm has been estimated at $125 billion. New Orleans accounted for $70 billion of that tab. New Orleans is one of the most strategically important cities in the United States. It's the last gateway of America's largest river, which drains 41 percent of the contiguous forty-eight states and carries 60 percent of our grain exports. The Port of New Orleans supports 150,000 jobs and sees 500 million tons of shipped goods pass through each year. With every resource imaginable devoted to restoring its damaged infrastructure, it still took a year before the port could handle prestorm levels of traffic.

There was a climate change lesson from Katrina, but it never registered, buried by the flood of other stories related to the flood— human interest stories; stories about the disproportionate damage to poorer parts of the community; stories about people trapped on their roofs, stories about looting, violence, and police overreaction; stories about pet rescues; stories about the diaspora that followed the storm. All were worthy of attention, but their sheer volume obscured the stark warning that the storm offered.

The climate message of Katrina was simple: preparations that enabled your city to withstand nature's blows throughout your hometown's history will not be sufficient for what nature has in store. It was a message that would be delivered to New York with Hurricane Sandy seven years later, and then to Houston with the arrival of Hurricane Harvey in 2017.

The combination of sea level rise, increased storm frequency, and increased storm intensity accompanying a warming globe has shown that coastal defenses built to withstand the storm surges of the past will not be adequate for tides and surges of the future. Nor is this message about weather preparedness confined to coastal cities. As many nations around the world are discovering, dams designed to contain floods based on historical data may not be adequate for the scale of floods that follow intensified rains. Harvey's 10-foot storm surge was exacerbated by an astounding 60 inches of rain that fell on the Houston area over a matter of days.

Katrina's most important message can be summed up in one word: thresholds. Had New Orleans's defenses been a bit higher, and their base better designed, the damages might have been measured in the millions rather than the tens of billions. A storm surge of 11 feet is 10 percent higher than one of 10 feet. If, to take a hypothetical example, levees protecting a city are built to contain a worst case of 10 feet, even a 5 percent increase in storm surge can up potential damages a thousandfold. These same thresholds apply to protection from winds, heat waves, droughts, and other weather-related phe-nomena. Beyond certain thresholds, an incremental increase can produce exponential increases in damage.

It's not just the United States that is learning that defenses built for the climate of the past ten thousand years may not be adequate in the new climate regime. The sea walls protecting the Chinese mega-city of Shanghai were built to defend the city from a once-in-a-two-hundred-year flood. The combination of sea level rise and subsidence will mean that the city can expect such floods as frequently as every twenty years by the second half of this century according to Gao Shu, a coastal defense expert, as reported by the Chinese publication *Sixth Tone*. A recent analysis undertaken by *Financial Times* estimated that sea level rise and attendant floods and tides could inflict nearly a trillion dollars in economic damage (relative to 2019 GDP) to the Shanghai area alone, with other low-lying coastal cities exposed to hundreds of billions of dollars in additional damage.

The expense for preventing this flooding is prohibitive. Gao, quoted in *Sixth Tone*, also noted that the price of bolstering these defenses rises exponentially if a wall has to be doubled in height from 1 to 2 meters. With some 15,000 kilometers of coast to protect with sea walls, the increased tab is beyond the reach of the Chinese gov-ernment. In a refreshing sign of a new way of approaching problems, China is looking at biological coastal defenses such as mangroves and marshes as a less expensive alternative.

Another threshold involves crops. Yields begin to drop precipi-tously as temperatures rise. A study of cereal crops in China showed that a 1 degree Celsius increase in nighttime temperatures led to a 10 percent drop in yields. Another couple of degrees increase in all-day temperatures and many crops won't grow at all.

Beyond certain thresholds, incremental increases lead to non-linear impacts that can be near impossible to predict. The failure of a rice crop might lead to starvation, bankruptcy for farmers and their suppliers, and a cascade of repercussions including social unrest and the fall of governments. This is exactly what happened to end the thirty-one-year Suharto regime in 1998 after a devastating El Niño ruined the rice crop.

These thresholds are real. Some, like the heat tolerances of cereal crops, are hard, natural limits, while others are more movable, involving the assumptions about weather made during the design of infrastructure. Whether natural or artificial, however, these thresholds make predicting the costs of climate change almost impossible because the cascade of repercussions of a threshold breached will be wildly unpredictable. I came upon this truth myself when I was working on a report attempting to predict the future costs of rapid climate change sponsored by the reinsurance industry and various international and public institutions.

The first decade of the new millennium unveiled other unwelcome aspects of climate change beyond accelerating sea level rise and the problem of thresholds. One was a lesson about the costs of what might be described as second-derivative impacts. In the 1990s, Paul Epstein, the World Health Organization, and many others warned about the relationship between climate change and disease in animals and plants. In the first decade of the 2000s, the United States got a vivid lesson in the relationship between climate change, plant disease, and wildfires.

Global warming sets the stage for larger and more intense fires in several ways. The heat dries out the vegetation and the soils. It also spurs population explosions for various bark beetles that can kill entire forests, thereby providing a ready supply of fuel waiting for a spark. A warming globe allows bark beetles to invade areas where cold temperatures previously kept them at bay, and warmer winters mean that fewer of these pests die off each year. And with extreme weather come more intense winds, which can make fires uncontrollable once they get going.

The record of the first decade of the millennium underscores the impact of these unholy synergies. In 2002, Arizona, which had suffered infestations of beetles killing ponderosa and piñon pines, suffered its worst fire in history. Colorado, also beleaguered by bark beetle infestations, suffered its worst fire that year. Two years later, the Taylor Complex Fire in eastern Alaska burned 1.3 million acres. That fire was the largest in North America for a ten-year stretch. Then in 2007, Georgia, Florida, and Utah also had their worst fires ever. In between there were plenty of second and third worsts in states such as California and Oregon, and right after the start of the next decade, Texas suffered its worst, while a new record for largest forest fire was set in Arizona, and a series of three fires set and subsequently broke the record for largest wildfires in New Mexico history.

As we now know, the frequency, scale, and intensity of wildfires have only increased in subsequent years as drought has become a recurrent feature of North American climate, not just in the West, but in the Midwest and South as well. The connection between climate change, beetle infestations, and wildfires was well covered by the press in the early 2000s, but any lessons that might have been drawn about the wisdom of building in potential fire zones went unheeded. Even as the incidence of fires rose, some of the most fire-prone areas of Arizona, California, and other states enjoyed a building boom, briefly interrupted not by worries about wildfires but by the housing bust that accompanied the crash of 2008.

The drumbeat of natural disasters continued to become louder and more urgent throughout the decade. A survey by Munich Re held that 2008 had a record forty category 5 natural disasters—in terms of financial and human impacts—only one of which, an earthquake in Japan, was not related to weather. As we will see, the public continued to ignore any connection to changing climate.

Not heeding nature's warning signals about climate change seemed to be the unifying theme of the first decade of the new millennium. People and businesses might be given a grudging pass for not taking climate change into account in the 1990s, particularly in the early part of the decade, as a picture of the warming world was emerging from background clutter amid plenty of competing claims on attention. By 2010, however, the message from nature was loud and clear: climate change was already here and promised to get more dangerous and expensive.

Twelve

THE SCIENCE
OF CLIMATE CHANGE
IN THE OUGHTS

For climate scientists, the first decade of the new millennium was marked by two radically divergent trends. There were many scientific breakthroughs and advances, but in terms of the public face of science, the decade was marked by backpedaling and bogus controversies. Some of the worst-case scenarios from the 1990s became the scientific consensus, but then two IPCC Summaries for Policymakers dialed back some of the extrapolations of earlier assessments. The mixed messages further added to public confusion about the imminence and seriousness of the threat of climate change, even as the scientific consensus solidified.

In 2004, Naomi Oreskes, a historian of science at Harvard, along with colleagues, published a sample review in *Science* of 928 peer-reviewed studies on climate change from a ten-year period between 1993 and 2003 to see whether any challenged the scientific consensus that human activities were causing climate to change. None did.

This was a stake in the heart of the denier community's biggest arguments for delay: that the science was unsettled and that there was active disagreement in the scientific community about the role of humans in changing climate. The deniers were not going to take this lying down.

First, a British scientist named Benny Peiser, a social anthropologist, repeated Oreskes's survey and claimed to have found thirty-four peer-reviewed studies challenging the consensus. Further inspection showed that the papers Peiser listed either didn't challenge the consensus after all or were letters or editorials that were not peer reviewed. Peiser retracted his criticism.

Not to be outdone, another energetic denier with the ten-dollar name of Viscount Monckton of Brenchley presented five articles challenging the consensus. These fizzled as well; some were reviews and not peer reviewed, while others either supported the consensus or had been amended to reflect the consensus.

This should have been the end of it, of course, but deniers have always had a prominent and friendly outlet in the editorial pages of *The Wall Street Journal,* and they were still giving space to Monckton's casuistry more than a decade after *Science* magazine put its weight behind the consensus. The sleight of hand is worth noting. As reported by the fact-checking organization Skeptical Science, Monckton compared papers that explicitly quantified the human contribution to climate change with all papers that used the words "global warming" or "global climate change."

This is a meaningless comparison because it implies that those papers on global warming that don't note the human contribution disagree with the consensus. As Dana Nuccitelli, an environmental scientist, pointed out in *The Guardian,* by this standard less than

1 percent of papers on astronomy accept the consensus that the earth rotates around the sun because they don't explicitly mention that fact. This kind of analysis would also provide supporting evidence for those who believe the earth is flat. Such examples utterly contradict any imputation that the editors of *The Wall Street Journal* or others who give space to such ludicrous analyses are acting in good faith.

There are, of course, real climate scientists who dispute the consensus. Roy Spencer of the University of Alabama is one. He teamed up with Joseph Bast, president of the Heartland Institute, a Koch brothers–financed astroturf outfit that serves as Grendel's Den of denialism. They published an op-ed in *The Wall Street Journal* attacking the notion of consensus. They didn't help their case when, as noted by Nuccitelli, they referenced a petition that could be signed by anyone with any sort of science degree, and whose signatories included one of the Spice Girls.

In any event, several subsequent studies all supported the consensus. One, conducted by Nuccitelli and colleagues, surveyed more than ten thousand scientists whose published papers referenced human-caused climate change. In this case, 98.4 percent supported the consensus. It's important to note that the consensus can be wrong—the earth rotates around the sun and not the other way around as was thought for thousands of years—but some arduously obtained scientific results are factual and enduring. Few, except for a couple of well-known climate deniers, would argue that smoking is not bad for the lungs. The scientific consensus on the human role in climate change is even more robust. Tragically, even today, most of the public doesn't know that.

Another noteworthy event in the jelling of that consensus was a publication in 2002 by the National Research Council of the National

Academy of Sciences, entitled *Abrupt Climate Change: Inevitable Surprises*. This came nine years after publication of the dramatic findings taken from the Greenland ice core projects that confirmed that climate had changed with extraordinary rapidity in the past. The span provides a data point about how long it takes for an emerging theory to become the consensus, even after it has been blessed by the most respected scientific journals. The National Academy report described the acknowledgment of rapid climate change as a "paradigm shift," lending its imprimatur to what almost all actively engaged scientists had accepted for years.

The simple logic chain following from this new paradigm should have been alarming to both policymakers and the public. If the past pattern of climate change had been characterized by rapid jumps, and if climate is already changing because of human inputs into the atmosphere, then it is likely that these changes will continue and come about rapidly.

The relatively ponderous path that the acceptance of rapid climate change took to go from the fringes to the conventional wisdom offered a cautionary tale for how long it might take for several other alarming discoveries about climate change to become consensus, much less have an impact on policymakers and the public. It took a couple of decades from the time Wallace Broecker and others first offered evidence that climate had changed rapidly in the past until the dramatic data from the dome of the Greenland Ice Sheet confirmed these past changes, and then another eight years before the majority of the scientific community accepted that this was the way climate shifts.

If rapid climate change posed a problem for scientists because of difficulties finding reliable proxies with sufficient resolution to reveal the rapid changes, another climate issue became even more controversial because of difficulties of measuring and interpreting how the various components of the phenomenon might interact. This was sea level rise, an issue with immediate pertinence to the roughly 300 million people who might be impacted in the coming decades.

Reasonably reliable measurements of global sea level only became possible in 1993 with the advent of satellite data. Previously, most measurements came from tidal gauges and buoys, which at best provided an incomplete picture. The likely contributions of glaciers and ice sheets were shrouded in even more uncertainty. When the first IPCC assessment came out in 1990, researchers were cautious about forecasts, asserting that they could only talk about possibilities, not predictions.

By the new millennium, however, many of these fuzzy areas had come into sharper focus. The melting of midlatitude and tropical glaciers was well underway and well documented. In 1991, retreating Swiss glaciers uncovered the preserved mummy of a Copper Age European hunter who died 5,300 years previously. Andean glacial melting uncovered plants that had been frozen for several thousand years, and tropical glaciers were well on their way to disappearing. Those atop Mount Kilimanjaro in Kenya shrank by 26 percent in the years between 2000 and 2007.

By 2000, the signal of sea level rise was unambiguous. For one thing, land-based glacial melting had accelerated sixfold between 1993 and the early 2010s according to a study out of the University of Bristol in the United Kingdom. The melting was having a discernible impact on sea levels.

After 2000, scientists quickly caught up to reality on several fronts, though the IPCC still lagged behind. If in the 1990s it was unclear whether the ice sheets would be a plus or a minus for sea level rise, by the middle of the decade it was clear that they would be contributing significantly to increases. If in 1990 there wasn't enough data to predict how much the permafrost might melt and how soon, by 2000, throughout the Arctic, buildings were tilting and roads buckling as the top layer of permafrost began to melt.

During that first decade of the new millennium, researchers concerned with the antipodes made significant progress in understanding the reaction of the ice sheets to climate change, and, in turn, the interconnections between ice sheets, glaciers, sea ice, and climate change. In 2001, *Science* published a satellite survey of two thousand glaciers that showed that most of them were shrinking. Surveys of sea ice showed clear evidence that the so-called ice minimums at the end of the Arctic summer in September were getting progressively smaller. As the summer extent of sea ice shrank, many scientists realized that the increased area of open water would have immediate impacts on Northern Hemisphere weather.

That's because whether a surface is reflective or absorbent—its albedo—can have a profound impact on the weather. While the white surface of sea ice reflects between 50 and 70 percent of incoming solar energy, the dark ocean reflects only about 6 percent of such energy and absorbs the rest. Thus, as sea ice retreats, the lower albedo of the open water amplifies the warming.

The shift in the oceans from the white surface of sea ice reflecting heat back into space to the dark surface of open water absorbing and releasing heat has a dramatic impact on climate. As open water expands, the oceans warm large expanses of Arctic air. With the warmer Arctic air, there is less contrast between the temperatures of the Far North and the midlatitudes. The sharper the contrast, the more vigorous the jet stream, but as temperatures in the Arctic warm and the contrast becomes less defined, the jet stream slows when, for instance, it moves over North America. In turn, this leads to exaggerated kinks in the high-altitude winds, which allow warm air to intrude northward and polar air to plunge south. The slowing of the jet stream means that weather patterns become extreme in other ways, notably in their persistence. Cold spells become longer and more intense, as do heat waves and droughts, and storms linger.

What happens in the Arctic doesn't stay in the Arctic. At the point in the 2000s when the retreat of sea ice became incontrovertible, the incredibly complex interactions between ice, the oceans, and the atmosphere in the Arctic were still on the frontiers of science. Indeed, this was perhaps the most extreme case of science being forced to develop explanations for the implications of global warming even as those implications were unfolding in real time.

In ecological terms a disaster was already underway as polar bears found themselves with shorter hunting seasons out on the ice, walruses were forced to use haulouts (places where they could rest and reproduce) on land rather than ice, which left them vulnerable to predators as the ice retreated farther and farther from shore, and countless other delicate balances, developed over millennia, suddenly were thrown out of kilter. In subsequent years the ecological toll has accelerated, marked by massive bird die-offs and tales of starving polar bears.

In the 2000s, the Arctic was already changing. I saw these changes for myself while reporting various articles for *Time*. In Yakutia I saw a baby mammoth that had been preserved for tens of thousands of years in the permafrost and was now in danger of decomposing as the warmth penetrated deeper into the ground above it. I also saw tilting buildings in Yakutsk, the capital, which became destabilized as the ground liquefied beneath them (in this case global warming might have been performing a public service, as many of these buildings were soulless legacies of the Stalin era). I visited a polar bear "jail" in Churchill on Hudson Bay, a facility where rangers put bears that strayed into town looking for food because they couldn't get enough sustenance from their normal hunting grounds as the warming Arctic melted the sea ice.

I concluded an article on climate changes in the Arctic by noting the irony that far northern defense installations on the front lines of the Cold War have been repurposed to study climate change:

> At the entrance to the Churchill Northern Studies Centre, a base for investigations of regional climate change, a rusting rocket is a mute reminder of the complex's earlier life as part of defenses against Soviet nuclear attack. That threat never materialized, and now, belatedly, scientists venture from the base to study a threat that has materialized but against which no adequate defense has been mounted. Despite the danger that climate change poses, the resources currently devoted to studying this problem—and combatting it—are inconsequential compared with the trillions spent during the cold war. Twenty years from now, we may wonder how we could have miscalculated which threat represented the greater peril.

Now it is twenty years later, and a good question for a public opinion expert might be how many people now see climate change as a bigger danger than the Soviets posed during the Cold War.

In that first decade of the 2000s, the Arctic saw the most dramatic changes in climate. Scientists, prominently Jennifer Francis, then at Rutgers and now at the Woodwell Climate Research Center, explored how these changes in circulatory patterns might impact the lower latitudes. She didn't have to wait long to find confirmation for most of her predictions as persistent cold spells (misleadingly dubbed "polar vortexes"), heat waves, and droughts became hallmarks of midlatitude life in North America during the very next decade, as did their connection to the extreme changes in the northern cryosphere.

Nor did scientists who were worried about the stability of the ice sheets have to wait long to see whether their shrinkage would contribute to sea level rise. A study done by NASA's Jet Propulsion Laboratory and the University of Kansas in 2006 found that the Greenland Ice Sheet was melting twice as fast as previously estimated based on satellite data going back to 1996. Later that year another study out of the University of Texas confirmed the acceleration, and subsequent studies out of the University of Bristol documented a further acceleration of the melting of the ice sheet in the years following 2006. The message of these studies was that Greenland was melting and the melting was accelerating.

As for Antarctica, melting and iceberg calving from the West Antarctic Ice Sheet were already contributing to sea level rise, although the question of how much would, in 2007, become a cause célèbre. Subsequent analysis showed that melting from all the ice sheets (Greenland and Antarctica) contributed between about 1.3 and 1.5 millimeters of sea level rise in 2006. This doesn't sound like much, but total sea level rise that year was roughly 3 millimeters, meaning that the ice sheets were contributing more than 40 percent of that increase.

This set the stage for some real drama over the next year as scientists and policymakers prepared to release the fourth IPCC assessment in 2007. By this point, most scientists agreed that sea level was rising about three times as fast as during most of the previous century, that the rate of sea level rise was accelerating, and that melting from the ice sheets was a significant factor in both the rate of sea level rise and the acceleration of the rate. On top of that, the consensus was that the rate of sea level rise would continue to accelerate; credible arguments that it would slow or level off were exceedingly difficult to find.

So how did the fourth assessment Summary for Policymakers treat sea level? It lowered its upper estimate of sea level rise from 88 centimeters in the previous assessment, published in 2001, to 59 centimeters. Both the 2001 and 2007 upper estimates for sea level were lower than the upper estimate of the first assessment in 1990. For the layman, the message of this consistent moderation of sea level rise estimates was that sea level was not going to be as severe a problem as initially thought and that, indeed, as scientists got better at estimating future sea level, the problem got less and less severe.

Needless to say, this was exactly the opposite of the message that the great preponderance of scientists wanted to convey, and so it's worth digging into how the IPCC got to this point.

It's no easy task to estimate future sea level rise given the wide variety of component inputs and uncertainties about future contributions from the ice sheets under various scenarios. To some degree the IPCC relied on what are called "process models." As leading climate modeler Stefan Rahmstorf explains it, process models "aim to simulate individual processes like thermal expansion or glacier melt." This seems like an entirely reasonable approach, but Rahmstorf and many other scientists felt that the models were not "mature." For one thing, at the time, they couldn't even reproduce the observed sea level rise since 1990, and the models underestimated actual sea level rise by 50 percent according to the German scientist.

To address this shortcoming, Rahmstorf developed a "semi-empirical" model for projections. Calibrating his model to past data, Rahmstorf's projections linked future sea level rise to global temperature data. He published his findings in *Science* in 2006, and they were vastly different from the IPCC projections. His upper case showed roughly twice as much sea level rise as the IPCC consensus. Though the paper elicited tremendous interest among scientists—Rahmstorf noted in the climate blog *RealClimate* that it was the second-most-cited paper of the ten thousand sea-level-related papers published between 2007 and 2013—the IPCC did not take semi-empirical models into consideration in assembling the chapter on sea level rise.

Why not?

It could have been a case of inherent scientific conservatism, though Rahmstorf has openly wondered whether the semi-empirical models would have been included had they projected *less* sea level rise than the process models. Naomi Oreskes wonders whether it's a case of torquing the results by omission. She and coauthors published a paper entitled "Climate Change Prediction: Erring on the Side of Least Drama?," where they argued that "scientists are biased not toward alarmism but rather the reverse: toward cautious estimates, where we define caution as erring on the side of less rather than more alarming predictions."

They go on to show that in case after case, the IPCC has been more willing to consider the outlier benign case over the outlier alarming case. Many others have pointed out that reality has been consistently worse than IPCC best estimates. A recent paper in *Nature Climate Change* shows that Greenland ice melt, for instance, has tracked or exceeded the IPCC's upper case (the widespread tendency is to say "worst case," but IPCC authors point out that they don't give a worst case, but rather an upper case within the bounds of probability).

This tendency toward conservatism has provided ammunition for the "nothing to see here" crowd. Patrick Michaels, who has made a career downplaying the threat of climate change, used data cherry-picked from the fourth assessment to attack Al Gore's book *An Inconvenient Truth*. Other climate deniers went over the report with a fine-tooth comb, looking for errors and feeding them to denier-friendly reporters such as Jonathan Leake. This led to a series of "gate" stories—"Amazongate," "Africagate," etc.—all of which turned out to be what I would call "vaporgates" upon examination.

At a time, then, when a clear message from scientists could have been "Climate changes are accelerating and are worse than we scientists anticipated," the muddled message to the public mediated by the deniers was a self-contradictory but effective "IPCC sees climate change as less of a threat" and "Don't trust the IPCC." Consequently, at a point when the public should have been hearing that scientific alarm about the scale and proximity of the climate peril was mounting, the message received was that scientists were still trying to sort things out.

The dustup over sea level rise and the contribution of the ice sheets at least involved scientific arguments, even if those arguments might have been tainted by political considerations. For the public, the three other scientific kerfuffles of the decade were pure disingenuous inventions of denier spinmeisters. The first two involved repercussions from the publication in the late 1990s of what was dubbed the "hockey stick" by Michael Mann and colleagues. Based on proxies, they showed that recent temperatures indicated that the planet was warmer than it had been in more than a thousand years and that the recent rises were so extreme that a graph of temperatures looked like a hockey stick.

The attacks were swift in coming. A Canadian mining consultant named Stephen McIntyre went after Mann, first for the legitimacy of the data, then for willfully misrepresenting the data. It bears noting that none of the challengers came from the paleoclimatology community or had any expertise in using proxies for reconstructing past climates (proxies, ranging from tree rings to lake bed sediments and ice cores, supplied most of Mann's data). McIntyre did find some small errors in proxy data that were not in the main body of Mann's analysis, but none of them would have affected the outcome, and McIntyre and his colleagues were swiftly rebutted for errors in their own methodology. That did not stop the fossil fuel lobby and their minions in Congress from seizing on this made-up "scandal."

McIntyre and a like-minded economist named Ross McKitrick were invited to meet with fossil fuel devotees such as Oklahoma senator James Inhofe, who once called climate change "the greatest hoax ever perpetrated on the American people." Rush Limbaugh blasted news of the supposed outrage, as did the editorial page of *The Wall Street Journal*. McIntyre and others besieged Mann and his colleagues with requests for "raw data" even when it was readily available on databases. The requests were so persistent that the scientists on the receiving end came to believe the demands were intended more to distract and annoy the scientists than for any other purpose.

As reported in *Mother Jones*, the University of East Anglia's Climatic Research Unit (CRU), a major keeper of temperature data, received fifty-eight Freedom of Information Act requests in just one week in July 2009.

But there was no scandal. The hockey stick was featured in the third assessment of the IPCC published in 2001. The National Academy of Sciences weighed in, supporting Mann's assertion of unprecedented warming. Subsequent analyses conducted by scores of scientists scattered over many institutions actually lengthened Mann's initial time period from one thousand to fourteen hundred years, and then to eleven thousand years. There were so many confirmations that Mann joked that rather than a hockey stick, they had a "hockey team." Projections of future temperatures make the hockey stick look even more frightening.

Step back and consider what the critics were arguing: they were saying that the entire paleoclimate community was mistaken and that the present warming was in accord with normal variations in climate over the past thousand years. In the early 2000s, there were countless other pieces of evidence that the warming was extraordinary— the melting of permafrost, the discovery of Ötzi the Iceman, a body entombed in ice 5,300 years ago but revealed as Swiss glaciers retreated, the march of insect-borne diseases into altitudes where they had never been before, the collapse of ice shelves that had been stable for thousands of years, the advancing date of spring and delayed autumns, the retreat and disappearance of glaciers that had been around for thousands of years, the opening of the Northwest Passage. Everywhere one looked there was evidence that climate changes were outside of normal bounds. And yet these critics were saying that if there was the tiniest error in one piece of data (and there wasn't), that all the other evidence didn't matter; we have to conclude that there was no warming. It was a classic case of "Who are you going to believe, us or your lying eyes?"

Nor was refutation of the accusations by every credible scientific body that looked at the issue sufficient to put the debate to rest. The argument trumpeted by the deniers was that there was a vast conspiracy of climate scientists to hype the threat because that's how they got government money. Putting aside the hilarious idea that the George W. Bush administration was going to channel money to scientists seeking supporting evidence of a climate crisis—that administration was more likely to subsidize critics of global warming—the very argument represented a profound misunderstanding of what motivates scientists. Scientists are born contrarians, and they make their reputation by challenging the consensus, not supporting it. Moreover, even then, there was direct evidence that the ones getting paid for their opinions were the ones throwing the accusations.

As the saying goes, when the facts aren't with you, challenge the messenger. First there were the vague allegations that all climate scientists were in it for the money. Then, when a hacker stole more than a thousand emails from CRU servers, there came accusations of outright fraud, that scientists were hiding data that contradicted the global warming thesis. Poring through the emails between various scientists trying to make sense of the proxy and instrumental data, critics jumped on words like "trick" and "decline" to show that scientists were conspiring the hide the awful truth that climate change was a big lie.

The usual crowd jumped on the "scandal." The BBC quoted the Saudi Arabian climate negotiator, hardly a neutral source as a representative of the world's most famous petro-state: "It appears from the details of the scandal that there is no relationship whatsoever between human activities and climate change." Other conservative outlets used such phrases as "the greatest deception in history."

I won't devote much space to Climategate because it turned out to be a scandal based on nothing. The "trick" turned out to be a standard statistical method, and the "decline" referred to a curious cool signal from tree rings during relatively recent periods when the instrumental record showed warming. Several subsequent investigations concluded that nothing in the stolen emails contradicted the published work of the scientists in any way whatsoever.

The phony scandal achieved its likely goal of making life miserable for the scientists involved. Never in his most paranoid dreams when he was a graduate student did Michael Mann think that his work on paleoclimate would lead to death threats, or that a reconstruction of past temperatures would lead Attorney General Ken Cuccinelli of Mann's home state, Virginia, to launch McCarthyesque investigations of his work (ultimately Mann left his position at the University of Virginia and moved to the friendlier state of Pennsylvania). Cuccinelli initiated a protracted legal battle ostensibly to see whether state money had been misspent on Mann's work. In his court filings, he continued to cite the stolen emails long after investigations had determined that they contained nothing improper. Ultimately, Cuccinelli lost in the state supreme court, but not before Mann and his defenders spent hundreds of thousands of dollars and countless hours trying to fend off what amounted to state-sponsored harassment of a distinguished scientist.

In a just world, Cuccinelli would have been driven off the public stage in shame. Unfortunately, in Trump's world, a despicable attack on a serious scientist made Cuccinelli their kind of guy. Trump would name him acting director of the United States Citizenship and Immigration Services, part of the Department of Homeland Security. And if someone is looking for a current "scandal," it turns out that Cuccinelli was appointed in contravention of federal law, and the Trump administration ultimately dropped its appeal to defend his appointment.

Despite the fact that every Climategate turned out to do nothing more than debase the significance of appending "-gate" to some alleged scandal, and despite the fact that the courts rebuked various efforts to harass working scientists, these fake scandals and lawsuits were fabulous successes for the hucksters promoting them. Six months after news of the stolen emails hit the headlines, a Gallup poll showed that the number of people who believed that global warming was an exaggerated threat jumped to 48 percent from 41 percent a year earlier. Polling in the United Kingdom showed similar results. And, as we shall see, the deniers were not done sowing confusion and doubt.

The first decade of the millennium turned out to be the warmest yet recorded. Scientifically, it turned out to be something of a schizophrenic decade. There were solid advances in the understanding of the components of sea level rise, particularly with regard to the role of the great ice sheets, but the most critical elements did not make it into the Summary for Policymakers, which, rightly or wrongly, was regarded as the scientific consensus. Consequently, if we take the IPCC fourth assessment as the public face of the state of the science in the decade, it looks as though the scientific clock ran backward during this time.

Thirteen

PUBLIC OPINION
IN THE OUGHTS:
A Climate Denier in
the White House

The U.S. public entered the 2000s with concerns other than climate change on its mind. If anything, the overall mood of the country was complacent. We were coming off the longest period of prosperity in American history. The U.S. budget was in surplus for the first time in memory, and while the dot-com stock bubble burst in March 2000, the real damage to equities was still in the future. Fast forward ten years to 2010, and the mood was anything but complacent. The events of 9/11 and the subsequent War on Terror, the Great Recession, the return of huge deficits and mass unemployment all contributed to a very different state of mind. One thing hadn't changed from the previous decade, though: global warming remained a matter of marginal concern for the broad public.

The dispiriting message: Americans were not seeing climate change as a threat during good times, when there were fewer crises to distract the public, nor were they seeing climate change as a threat in bad times, when the public was more attuned to other potential dangers. One reason was the persistent efforts of the deniers to confuse the public. Another major factor: few people in positions of power were asking the public to see climate change as a problem.

A meta factor that has only become more evident over the past twenty years has been an increasing divergence of the interests of the public and matters of public interest. Contributing to this has been the partisanship and political demonization that accelerated with Newt Gingrich's takeover of Congress in 1994, accelerated during the Clinton impeachment hearings, and has been fueled subsequently by the advent of the Tea Party and the election of Donald Trump. Part and parcel of this syndrome has been the mockery of expertise, an infantile trend of "owning the libs," and a general trivialization of political discourse.

The decade began with a U.S. presidential election. One of the candidates, Vice President Al Gore, had been a vocal internal advocate for action on climate change during his two terms in the Executive Office Building. During the campaign? Silence.

He talked about Social Security a lot. He talked about jobs. He talked about a lot of things, but he spoke very little about climate change.

For those of us who were looking for a leader to raise the profile of the issue, this was disappointing but not necessarily surprising. Gore's electoral strategy focused on holding on to the rust-belt states—Illinois, Michigan, Ohio, West Virginia—and in the minds of his strategists, this meant, among other things, not alienating the big unions such as the United Auto Workers and the United Mine Workers. I felt that this strategy made little sense, since his position on global warming and environmental issues was well known. Hiding those positions during a campaign was not going to fool anyone and would only underscore his reputation as a slick politician.

As for the judgment of his pollsters and strategists that climate change was not an issue for the voters he needed? Unfortunately, they were probably dead right. No one who cared about the issue was going to vote for George W. Bush, an oil guy from an oil state. And certainly anyone concerned about the issue knew that, once elected, Gore would have been a far better advocate for action than Bush.

But would it have hurt Gore to identify climate change as a critical issue facing our planet? He had the credibility of the world's most august scientific institutions behind him, and he could have made the case that dealing with the threat could be an economic plus, while the consequences of climate change could result in economic disaster. He also could have reassured the unions that he would continue the policies that contributed to the unparalleled prosperity of the Clinton years. In any event, where were the unions going to go? It defied logic that they would turn out for Bush, who came from the most anti-union faction of the Republican Party.

As it turned out, Gore won the popular vote by a substantial margin but lost rust-belt states Ohio and West Virginia. He also lost (but, infamously, may have won) Florida. The deciding factor had nothing to do with climate change or Gore's environmental policies. Rather it was Ralph Nader, whose presence in the race siphoned off precious votes in two critical states.

Sometime after the election, I caught up with Gore backstage after he gave a talk at New York's Beacon Theater. I asked why he hadn't made global warming a part of his campaign. He was quite candid, saying, "What about the UAW?"

Soon after Bush entered office, his administration rammed through a massive tax cut ("We won. It's our due," Vice President Cheney famously remarked at the time), and deficits returned with a vengeance. Bush also vigorously started appointing fossil fuel and industry lobbyists to positions relating to climate change and environment, a fox-guarding-the-chicken-coop policy almost as egregious as Donald Trump's later approach.

If the Clinton years represented a lost opportunity on climate change because of political timidity, during the George W. Bush years, the United States actually retreated from confronting the problem even as the evidence mounted that the effects of climate change were already here in the form of wildfires, floods, and other expensive natural disasters.

Within two hours of Bush being sworn in, his chief of staff, Andrew Card, issued a memo placing a hold on new regulations, withdrawing regulations not yet published in the *Federal Register*, and postponing regulations that were not yet in effect. This included all environmental regulations. One of these was a Clinton rule on energy efficiency—a key tool to reduce carbon emissions. Ultimately, this rollback was blocked in the courts, but the fight set the tone for the administration: instead of looking for ways to curb greenhouse gas emissions, advocates for action on climate change found themselves desperately trying to hold on to meager climate initiatives implemented by prior administrations.

As for the quality of Bush appointments to sensitive environmental positions, consider Steven Griles, a lobbyist for the coal industry, who was named deputy secretary of the interior, the cabinet department with responsibility for drilling and mining on public lands. During his short time in office, Griles had a "come on in, take a seat, and can I get

you a cup of coffee?" policy for mining interests, with visits often bro-
kered by legendary fixer Jack Abramoff. Griles lied to investigators
about these ties, ultimately pleaded guilty to obstruction of justice,
and was sentenced to ten months in prison.

Another typical appointment was Jeffrey Holmstead, who put
aside his work as an attorney representing the nation's worst air polluters to take charge of the division of EPA responsible for regulating
air pollution. Bush's head of NASA for his second term was Michael
Griffin, who tried to muzzle James Hansen, the agency's most celebrated expert on climate change. During an interview with NPR, he
was at best blasé about global warming, doubting that "it was a problem we must wrestle with."

Perhaps his worst appointment was Philip Cooney, named as
head of the White House Council on Environmental Quality. A former lobbyist for the American Petroleum Institute, Cooney maintained his close ties to the fossil fuel industry in his new position as
chief White House advocate for the environment. Cooney went beyond muzzling experts (though he did that too) to actually editing the
work produced by them. Specifically, he made 364 editorial changes
to the administration's plan to deal with climate change, most of
which, a subsequent House investigation revealed, served to amplify
scientific uncertainty and downplay the human contribution to climate
change. Here's a summary sentence from that House report: "The
Bush Administration has engaged in a systematic effort to manipu-
late climate change science and mislead policymakers." The administration was directly contradicting the scientific consensus on the key
elements of climate change. Once the scandal was exposed, Cooney
was forced to resign.

It was no surprise that the administration refused to abide by the

Kyoto treaty on climate change. During one interview, Bush said that Kyoto would have "wrecked" our economy. The 191 countries that ratified the treaty clearly didn't feel that way about their own economies. Bush did have half a point in arguing that the treaty did not require emissions reductions from China and India, which, even then, were on their way to becoming the world's biggest emitters of greenhouse gases. Still, given the fossil fuel credentials of all his appointments, it stretches credulity to contend that citing China and India was anything more than a convenient excuse.

President Bush clearly wanted nothing to do with climate change, but an opportunity to force him to confront the issue arose in 2004 during his second presidential election. The Democratic candidate, Senator John Kerry, had excellent credentials on the issue and a deep bench of advisers conversant with all the ways it might be presented to voters as both a threat to our way of life and an opportunity to spur economic growth. It was an opportunity not taken.

Kerry focused on an increasingly unpopular war in Iraq, on jobs, and on affordable health care, and while he gave a couple of speeches on climate change, at no point was it the issue of the day. A campaign team will prep the major media with the issues that the candidate will be focusing on in the coming days. If climate change is not on the list or near the top of the list, the political reporters are not going to focus on the issue. I had numerous conversations with various Kerry campaign aides during that election, and climate was never near the top.

As was the case with the Gore campaign, the real damage of the neglect of the issue was to relegate it to second-class status in the mainstream press. During a presidential election, the political reporters dominate the space in a newspaper or newscast. If the Kerry campaign had featured global warming as an issue, the political press would have covered it, particularly because there was such a sharp contrast between the seriousness with which Kerry took the issue and Bush's dismissive attitude. And if the political press had covered the issue, it would have told the public that this was serious. Instead, coverage of climate change was mostly buried during the months that politics dominated the news agenda.

There's a bit of a catch-22 in this. Kerry didn't make climate change a major issue during the campaign because his advisers told him that it would not help him get elected. They were right, but one reason that it was not a major concern for voters was that no presidential candidate was making an eloquent case that it should be a major concern.*

As it turned out, through most of the decade, news coverage of climate change was at best intermittent. Without the rocket fuel of the issue becoming a major issue in a presidential campaign, much of the coverage it did receive had to do with the science or the faux outrage from the right over various trumped-up scandals like Climategate. More in-depth coverage was left to books, and there were plenty to choose from.

In 2006, Al Gore published *An Inconvenient Truth*, accompanied by a documentary. The film became a sensation, winning two Academy Awards, and the book became a best seller.

There were several other books that year that tackled the issue from various angles. My book, *Winds of Change*, explored the role of

---

* It should be noted that at a Brookings Institution briefing after the election, I asked Kerry why his campaign didn't focus on climate change. He took vehement issue with the question (I'm guessing he'd been briefed that it might come up because I had been talking with campaign staff). Frankly, these protestations didn't wash then and don't wash now. I've yet to read one retrospective on his campaign arguing that he made climate change a prominent issue when he ran for president.

climate in the rise and fall of civilizations. Elizabeth Kolbert published her reporting on climate change in the high latitudes in *Field Notes from a Catastrophe*, and Australian scientist Tim Flannery's *The Weather Makers* came out as well. All these books and several more received broad coverage and prominent book reviews. The following year, Al Gore and the IPCC jointly shared the Nobel Peace Prize for their work on climate change, which is about as much attention as one can get. Did it make a difference in moving public opinion? Unfortunately, it did not.

The problem was not that the public was uninformed, although they were, but rather that the issue had become partisan after the Clinton impeachment, and facts and risks were ignored or adjusted depending on the source. For the most part Democrats understood that climate change was a problem in the 2000s, though not enough felt the threat sufficiently imminent to vote on the issue. Most Republicans, however, saw it as a Democratic issue pushed by bleeding hearts and Greens.

Moreover, the president of the United States, George W. Bush, dismissed the threat, so that scoffing at global warming became something of a loyalty test for Republicans. This led to the bizarre situation in which some of the states most at risk for climate change had governors who dismissed, derided, and ducked the issue. Then governor of Florida Jeb Bush helped pioneer the feint "I'm not a scientist" when asked about the issue. He also questioned whether scientists agreed that it was caused by humans (he's since changed his public views). Subsequent Republican governors Rick Scott and Ron DeSantis only grew more bombastic about the issue even as global warming inflicted economic harm on their state. Texas governor Rick Perry dismissed global warming as a "phony mess." Sonny

Perdue, governor of Georgia through most of the 2000s, was steadfast in his dismissal of the problem and maintained this position when he became Trump's secretary of agriculture.

I suspect that all the politicians who dismiss global warming despite its threat to the economies of their states made a political calculation that their voters cared more about their leaders hewing to the party line than acknowledging climate change as a real and present danger.

It's tempting to explain this as an issue of education and, in fact, Pew research has shown that around the world, more-educated people are more concerned about climate change than the lesser educated. But that same study by Pew in 2019 showed that four of the ten most highly educated nations on earth—Australia, Canada, the United States, and Israel—had a relatively low percentage of people who cited climate change as a major concern. As Anthony Leiserowitz pointed out, three of those countries are English-speaking former frontier colonies, have major interests in fossil fuels, and also have an active denier movement (as Australia's former prime minister Malcolm Turnbull put it in a 2020 debate on the Murdoch empire's impact on global warming, they "have turned this issue of physics into an issue of values or identity"). As for Israel, the low level of concern about global warming (the lowest of thirty-eight nations in the Pew study) is a bit of a mystery, since Israel could be hurt by a number of climate change impacts such as sea level rise, stresses on fresh water, and temperature rises that have already made parts of the Middle East uninhabitable. It could be that Israelis have plenty on their minds with hostile neighbors threatening war on a daily basis, or it could be that Israelis feel that their extraordinary technological expertise will keep them safe. They might be mindful, however, that climate change will impact their neighbors who don't have Israeli expertise, and that could be destabilizing for Israel as well.

Just three weeks after President Bush was inaugurated for his second term in 2005, the Kyoto Protocol came into force—on paper. In fact, there was very little force behind it. Thanks to Senate Resolution 98, introduced by Senators Byrd of West Virginia and Hagel of Nebraska and passed 95 to 0 in 1997, there was no chance that the United States would ratify the treaty, because it did not impose emissions reductions on China, India, and other developing nations.

Here's what was supposed to happen with the Kyoto agreement: by 2012, the developed nations were supposed to lower their greenhouse gas emissions to at least 4 percent under 1990 levels. Looked at from this narrow perspective, the treaty was a success because thirty-six nations handily beat that goal (although most of the success came from the modernization of the former Soviet states and the economic pullback during the Great Recession of 2008–2009, which cut global emissions by 14 percent).

If the goal, however, was to stabilize international GHG emissions, the treaty was an utter failure. Globally, emissions from combustion grew by 58 percent between 1990 and 2012 (the date chosen for the first test of the treaty's performance). Much of that increase came from China, which, for instance, accounted for 71 percent of the increase in emissions in 2012. That same year, U.S. emissions dropped by 4 percent.

Clearly, the greenhouse gas situation would be better today had China and India joined a treaty requiring emission reductions. A combination of incentives and disincentives might have guided the two giant nations toward development paths less dependent on fossil fuels. This does not let the Bush administration off the hook. What incentive did China have to agree to cut emissions when the president of the world's largest economy was actively disputing that climate change posed any threat at all—and also refusing to abide by the treaty?

The first decade of the twenty-first century ended as the hottest decade since reliable global temperature records began in the 1800s. And yet the decade ended with most of the public still thinking that the impacts of climate change were speculative and way off in the future and that the role of humans in climate change was still a matter of active debate among scientists. Following the end of the decade, Anthony Leiserowitz offered a depressing accounting of change in public opinion. He found that a group he labeled "dismissive" (those who believed that global warming was a hoax) more than doubled between 2008 and 2010 to 16 percent of the public, and the segment of the public he called "alarmed" dropped from 18 percent to 10 percent. He attributed this to a "perfect storm" of economic factors, disgust with Washington, and the efforts of deniers. Thus, the decade ended with less public alarm and more doubt about climate change, even as most of the major scientific questions about whether global warming posed a threat had been long since settled.

Fourteen

BUSINESS AND FINANCE
IN THE OUGHTS:
Stirrings of Change

While the oughts were another lost decade in terms of mobilizing public opinion, there were a few positive developments on the business and finance front—though not in the United States. The most significant was the dramatic expansion of feed-in tariffs (the mechanism pioneered in Europe that provided a guaranteed return to investors in renewables). Making renewables financeable turned out to be the key to dramatic expansion of solar, wind, and other alternatives in a host of nations. The passage of Germany's renewable energy act in 2000 provided a template adopted by many other nations for financing renewable projects in a way that was fair to both investors and consumers. First other European nations got on board, including France, Italy, Spain, Portugal, and the Czech Republic, then China, India, South Africa, and many other nations. Today, some fifty countries (and many individual states in the United States and provinces in Canada) have some form of feed-in tariff.

Another significant breakthrough came, surprisingly, from a collaboration of a physicist and an evolutionary biologist, and it didn't generate the attention it deserved. In 2004, Princeton University professors Rob Socolow and Stephen Pacala published a paper in *Science* in which they provided an elegant, commonsense framework for understanding the task of stabilizing greenhouse gas emissions, as well as a dramatic way of framing the costs of delay. The article, entitled "Stabilization Wedges: Solving the Climate Problem for the Next 50 Years with Current Technologies," broke the problem down. They showed how much global emissions had to be reduced in order to stabilize greenhouse gases over the next fifty years and then offered a path to achieve those reductions. They identified fifteen "wedges," each of which would reduce emissions by 1 gigaton. They called them wedges because the longer one delayed in implementing the reductions, the more was required to achieve stabilization. The wedge concept offered an easy way to visualize the costs of delay, because with the passage of time more would have to be done in a shorter period to achieve the goals of any single wedge. If time was the horizontal axis, then the vertical axis—what would have to be done—would get thicker with each year of delay, creating a wedge. In 2004, they calculated that the world needed to reduce emissions by 7 billion tons each year, so that any seven of the fifteen wedges would achieve the goal.

When I interviewed him, Pacala said that "we wrote that article in *Science* for an audience of one." He'd been infuriated when Bush's energy secretary, Spencer Abraham, said that to deal with climate change, we needed a breakthrough as profound as Michael Faraday's invention of the electric motor. Pacala and Socolow thought that was nonsense and set about trying to prove it.

The wedges ranged from doubling fuel economy for 2 billion cars, to increasing wind-generated electricity a hundredfold (relative to what was produced in 2004), to eliminating tropical deforestation. The World Resources Institute teamed up with Goldman Sachs to try to put some flesh on one of those wedges—increasing solar power. Unfortunately, they focused on concentrated solar power (the use of mirrors to focus sunlight and generate heat, which then drives a generator) at a time when plummeting solar cell prices made concentrated solar less competitive. This illustrates one of the pitfalls of technological wedges—namely that as pioneering technologies rapidly expand their footprint, it isn't obvious which form of the technology will win out.

Once Socolow and Pacala offered their architecture for how to achieve stabilization with existing technologies, the door was open for others to offer their own wedges. And they have; among those advanced have been carbon-tax wedges and meat consumption–reduction wedges. The Carbon Mitigation Initiative even has a game where teams can come up with their own strategy using their evolving list of wedges.

The good news is that in some cases the world is ahead of schedule in filling in some of those wedges, in some cases way ahead. While the authors estimated in 2004 that the world would need to increase wind-generated electricity by a hundredfold by 2050, wind capacity has since increased so rapidly that what remains to be achieved is only a tenfold increase by 2060. That is completely doable, as we've had more than a tenfold increase in just eighteen years. Solar power too has spread more rapidly than expected in 2004. While reducing emissions by a gigaton in 2004 required seven hundred times installed solar capacity, the figure is now a hundredfold increase in what is currently online. In 2004, calculations were that solar would have to grow by about 14 percent a year to fill that wedge. In 2017, photovoltaic installations grew by 34 percent.

Pacala says that the advances in wind and solar have been far more dramatic than they dreamed possible in 2004, and that those two renewables could account for much more than two wedges by 2050, even without further technological breakthroughs. "We never expected in 2004, that fifteen years later, they would be the cheapest forms of electrical energy." He attributes the extraordinary progress to market competition and subsidies. "Even in the past ten years, the costs of wind and solar have come down between four and five times. The success of wind and solar should be a conservative cause célèbre," he remarked.

Pacala's optimism on solar has been backed up by the International Energy Agency (IEA), which has found that the cost of solar has consistently dropped faster than even the agency's optimistic projections. In its *World Energy Outlook 2020*, the IEA argued that solar had become "the cheapest source of electricity in history." In just the year since the 2019 *Outlook*, solar had become 20 to 50 percent less expensive.

In fact, the decarbonization of electricity in the United States has dramatically exceeded expectations of fifteen years ago. A Lawrence Berkeley National Laboratory study, *Halfway to Zero: Progress Towards a Carbon-Free Power Sector*, led by Ryan Wiser, looked back at 2005 expectations for future electricity production and compared them with what actually happened as of 2020. As the study's title asserts, they found that in 2020, direct power sector emissions were half of what was expected by experts in 2005, just fifteen years earlier. One reason was the shale revolution, which made natural gas plentiful and dirt cheap. Another was the growth of renewables, which exceeded expectations by about 90 percent. The biggest reason for the decline, however, was an extraordinary decrease in power produced by coal. Instead of rising by 480 billion kilowatt-hours, coal power plummeted by 1.720 trillion kilowatt-hours, a nearly two-thirds drop!

Pacala cited the degree to which natural gas has displaced coal as a positive surprise—though he admits that too much escaped methane during drilling could nullify the advantage of natural gas (methane has twenty-eight times the greenhouse gas potential of carbon dioxide). Apart from the shifts in the power sector, another positive surprise is the surge in the shift to electric vehicles. "Talk in 2004 was all about increasing vehicle mileage," he said. "Nowhere were electric cars in the conversation. For one thing, we lacked the capacity to deal with millions of distributed sources."

The enabling technology, he notes, came from an unexpected quarter. The explosive growth of smartphones drove innovation in battery technology. Lithium-ion batteries could be scaled up, and competition among electric vehicle makers has dramatically expanded the range of the vehicles, lowered the costs of the batteries, and extended their life.

The surprising power of innovation and technological transformation has caused Pacala to up his goals. "The natural surprises have been on the ugly side," he notes. "The technological surprises have been on the bright side." Back in 2004, he felt it would be a home run to simply freeze emissions at 2004 levels by 2050. Now, with climate change already hitting hard, he acknowledges that this is not enough.

"Back in 2004, no one would have predicted that across all industrial nations, emissions are declining in every respect," he noted. But the problem is that emissions in developing nations are still rising. The new goal is net-zero emissions, and he feels that achieving that goal by 2060 is entirely feasible.

While the concept of wedges offers a path forward, it also offers a warning. Recall that each of the wedges gets fatter over time, reflecting the fact that as each year passes without stabilization, more greenhouse gases accumulate and more future emissions must be cut in a shorter time. Because greenhouse gas emissions have increased since 2004, the amount that must be cut to achieve stabilization has increased from 7 gigatons a year to 8 gigatons a year. With global emissions still growing at 2 percent a year (and expected to grow by 5 percent in 2021 as the world recovers from COVID-related shutdowns), that number will continue to rise, making it ever more difficult to stabilize emissions at a level that doesn't entail catastrophic warming.

Another accelerating trend during the oughts was an increasing divergence between the United States and the rest of the world with regard to action on climate change. Perhaps nowhere was this more striking than in policies relating to coal, the most egregious contributor to greenhouse gas emissions. Shortly after his inauguration, George W. Bush launched a National Energy Policy Plan under the leadership of Vice President Dick Cheney. It called for "regulatory certainty" for coal, a code word for a rollback of regulations. Bush followed that up by launching a Clear Skies Initiative (yes, there's a long history of Orwellian naming habits in the White House), which actually did roll back regulations. In response, coal production trended up throughout the Bush years.

The rise in coal production came to a screeching halt by the end of 2008 and then plummeted. The reason had nothing to do with the readoption of regulations or any "war on coal" instituted by the new Obama administration. Rather, a flood of cheap natural gas and (a few years later) the extraordinary rise in wind power made coal uneconomical for electricity generation. Support for this conclusion came from the Trump administration, perhaps the most pro-coal government anywhere on earth, anytime in history. Despite efforts that extended to attempts to rig markets to force generating companies to buy from the dirtiest, least efficient plants, coal use continued to fall steadily during his administration. That said, there is no doubt that it would have fallen faster but for the Trump administration's desperate measures, and it would have been even further along the curve to oblivion were it not for the George W. Bush administration's attempts to bolster coal use.

For a picture of what might have happened in the United States were it not for four coal-friendly administrations, consider the path taken by Germany, another economic powerhouse with major coal reserves. Since the adoption of the Renewable Energy Act, German coal use has seen a steady decline. More to the point, the country beat the goal of a 40 percent cut in greenhouse gas emissions by 2020. By contrast (and only due to COVID-19), the United States ended 2020 with greenhouse gas emissions slightly lower than 1990 levels. German achievements have come about despite having cut back on its reliance on nuclear power, which put an added burden on other sources to make up the difference.

The dramatic gap between the United States and Germany was the result of an enormous surge in renewables in Germany, principally wind and solar. Between 2000 and 2020, German power production from nonhydro renewables has grown more than tenfold and now accounts for roughly 40 percent of power generation. United States reliance on renewables has also grown, but nonhydro electrical production is still only 11 percent of power generation.

It must be stipulated that Germany and the United States are not perfectly comparable. Germany had the benefit of the modernization of the former East German territory after the Berlin Wall came down. On the other hand, the United States has vastly more land at latitudes better suited for solar than Germany, as well as vastly more potential for wind energy. Moreover, while Germany has dramatically outperformed the United States in terms of cutting emissions, it only ranked twenty-third on the Climate Protection Index, a measure of emissions reduction performance put together by a consortium of climate groups. Sweden ranked fourth (the index leaves the top three spots open because the judges feel no country has done enough to avoid dramatic warming), and the United States ranked dead last of the sixty-one nations assessed, right behind Saudi Arabia.

Two terms of George W. Bush inflicted both real and intangible damage on action on climate change. The tangible damage came from the increase in fossil fuel emissions at a time when other developed nations were cutting. Incalculable damage was done by having the administration of the world's largest economy and greenhouse gas emitter spend eight years actively sabotaging efforts to combat global warming. GOP mockery of the threat gave cover for politicians in other countries to drag their feet. There is no doubt that had Al Gore been president during those eight years, the atmosphere would be less burdened with greenhouse gases today.

The United States is not a monolith, of course. Even as the White House undermined action, states and cities launched their own initiatives. California, then the seventh-largest economy on the planet (and now the fifth), passed a Global Warming Solutions Act in 2006, setting a goal of reducing the state's emissions by 40 percent by 2030. Other states and dozens of cities adopted decarbonizing measures. Unfortunately, in global terms the absence of support from the federal government was the bell that rang the loudest.

The indictment of the United States in no way blesses the actions of the other developed nations. For instance, while the European Union has done a better job of cutting its emissions than the United States has, those nations "import" a significant amount of emissions from developing countries. When an Italian buys a toy made in China, the making of which produces a pound of CO_2, the greenhouse gas emission is produced in China, but its production was stimulated by demand in Italy. In this fashion the European Union has "offshored" its emissions. It looks good on paper, but the atmosphere has significant additional greenhouse gases created by the European Union that is not on the European Union's books. A study that was led by system scientist Steve Davis that was published in the *Proceedings of the National Academy of Sciences* in 2010 found that Austria, France, Sweden, Switzerland, and the United Kingdom imported about one-third of their emissions in that fashion.

There were other accounting tricks. In the oughts, a good portion of purported emissions cuts claimed by the European Union, for instance, came from a misbegotten project that was a brainchild of the Kyoto Protocol. To incentivize developing nations to industrialize while reducing emissions, the treaty had a Clean Development Mechanism (CDM) through which industries could pay for projects in other countries that would show marginal improvements in emissions and then use those credits to offset their own emissions. The idea seemed ingenious—a factory that might spend $100 million reducing carbon emissions might get more bang for their buck in terms of emissions by spending the money on a solar power project in Ecuador.

There were two problems from the get-go. One was that it was a zero-sum game; even if it worked properly, the CDM didn't reduce emissions, it just didn't increase them. The second was that it didn't work properly: much of the money went to those who figured out how to game the system, and some of the money actually increased emissions. Credits, for instance, went to landfills to capture methane even though some of the methane-capture projects had been up and running for more than a decade. In one case, a major credit went to build a coal-fired power plant in Gujarat that became a significant source of greenhouse gas emissions in India.

One problem was that for the CDM to work as intended, the sellers and buyers of the credits actually had to care about reducing greenhouse gas emissions. A significant number of participants just wanted either the credits or money and couldn't care less about saving the planet from climate change. The CDM also contained perverse incentives in the sense that it undermined efforts a nation might take to reduce emissions without being paid to do so. Why spend money to reduce emissions if you could get paid to do it?

The cap-and-trade system of the CDM is one attempt to put a price on carbon, and many were launched in the first decade of the new millennium. In 2005, the European Union started its own trading system to limit carbon, which is now the largest such system in the world. Consortiums of states in the United States, including California, provinces in Canada, states in Australia, and various countries around the world have also tried to put a price on carbon. The overarching principle is the simple concept of "polluter pays." It was used successfully in the early 1990s to reduce acid rain by setting up a market to put a price on sulfur dioxide emissions. The carbon market has expanded wildly, with $215 billion in carbon trading carried out in 2019 according to the market data firm Refinitiv.

The problem is that these programs have done little to reduce emissions. Many price carbon too cheaply, which is why cap-and-trade schemes are often popular with the big carbon emitters. In many cases, the supply of carbon permits is greater than the demand, which drives down prices. A survey of carbon pricing schemes by Matto Mildenberger and Leah Stokes for the *Boston Review* notes that half the world's carbon prices are less than ten dollars a ton, between a fourth and an eighth of what the World Bank assessed as necessary to meet the mild terms of the Paris Agreement and a tiny fraction of what would be necessary to actually prevent global temperatures from rising beyond 2 degrees Celsius.

Polluters also like carbon trading schemes because many are structured in such a way that carbon prices directly hit consumers, which makes it easy to launch slick campaigns arguing that climate change programs are vanity projects of insufferable elites that hurt the little people. In many cases, carbon trading schemes are a win-win-win for polluters. If they have to live with them, as in California, their lobbyists can make sure they're not too pricey. Also, polluters can use the threat of higher prices to rouse the working public, and quite often, if the emitting industry is sufficiently critical to an economy, they can stop them altogether. Or polluters can get an exemption, as was the case in New Zealand, where agriculture was exempt from the carbon trading scheme even though the sector accounted for half of the country's greenhouse gas emissions. To make things worse, state and regional trading schemes often pop up in areas that already have strong climate action programs and not where they might do the most good.

James Hansen recognized these flaws early on and testified about them before Congress. His preference was for a carbon tax. It's more difficult to game and could be targeted, as a small number of giant businesses account for a significant percentage of emissions. For instance, Secunda, a coal-powered chemical plant owned by Sasol in South Africa, has been named as the largest single-site emitter of greenhouse gases in the world. As reported by Bloomberg, its 56.5 million tons of greenhouse gases are more than the annual emissions of countries such as Norway and Portugal. The *Carbon Majors Report* asserts that just one hundred companies are responsible for 71 percent of greenhouse gas emissions since 1988.

Unfortunately, it's also easy to rally people against a new tax, and that's exactly what has happened when a carbon tax has been proposed over the past decades. In fact, opposition to a carbon tax in Australia helped propel Tony Abbott to become prime minister in 2013. (Abbott has said that he thought climate change would turn out to be a good thing. He made these remarks before a substantial part of his country was incinerated by wildfires in 2018.) The fact that a carbon tax can be recycled as a dividend to relieve the burden on the poor and middle class gets lost in the crude bumper-sticker messaging that befogs the discussion of any new tax proposal.

As noted, the insurance industry had the potential to become the most colossal financial force for change with regard to climate. Yet the industry was not a force for change in the 1990s and still wasn't in the oughts. One reason was their aforementioned ingenuity at spreading risk, while another was that in the regions most exposed to climate risk, the state stepped in to socialize the risk, albeit in a camouflaged way. Exhibit A is the state of Florida.

It's been argued that most of the growth in Florida since 1960 could be attributed to two factors: the spread of air-conditioning and the availability of affordable property insurance. Whatever the reason, the growth has been spectacular. By the end of the decade roughly 80 percent of insured real estate in Florida was on or near a coast. Notably, much of this growth occurred after Hurricane Andrew, and even after the extreme hurricane seasons of 2004 and 2005. (The first year set the record as the costliest season in U.S. history, and 2005 surpassed that record.)

In a world of functioning risk markets, insurance costs and bigger exclusions should have begun an orderly movement of people away from the coasts. The retreat should have extended beyond Florida to Alabama, Georgia, South Carolina, and other regions in harm's way. Instead, the opposite happened.

Reinsurers did pull back from Florida and other at-risk markets after the 2004–2005 hurricane-related losses. This threatened to precipitate a spiral in which either increasing insurance costs or unavailability of wind insurance would cause real estate prices to crash and banks to go under. The state dodged this bullet thanks to a plethora of state and federal programs that stepped in to fill the gap with underpriced risk insurance (a real estate crash did happen in Florida in 2008, but the cause was the collapse of the mortgage market, not hurricane risk).

Florida's backstops date back to 1970, and there are now three Florida programs plus federal flood insurance assuming risks that the private market won't insure or supporting policies that would otherwise be unaffordable. Because these programs subsidize risks private insurers will not, they more resemble public spending programs than insurance. The most dramatic example of this is the National Flood Insurance Program—NFIP—which has been in deficit since Hurricane Katrina. In 2016, Congress increased its debt by $16 billion, and then in 2018 canceled $16 billion in debt owed to the government (actually, to taxpayers). This is not a business model any private insurance company could emulate, as private companies do not have the option of canceling debt outside of bankruptcy or restructuring.

Florida's most recent program is the Citizens Property Insurance Corporation established by the Florida legislature in 2002 by rolling two existing backstops into a nonprofit insurer of last resort with a cap on the rates it could charge. It quickly became the largest insurer in the state, which tells you all you need to know about whether other insurers think that the rates charged are sufficient to sustain a viable business. Big insurers such as Allstate and State Farm pulled back after the losses of 2004 and 2005, and by 2012, Citizens was insuring 1.4 million properties. The private market was largely given over to smaller undercapitalized insurers.

Risk can be ignored, it can be spread, and it can be shifted, but if it is real, it cannot be destroyed. Should some major storm bankrupt the pool, the state was empowered to cover the losses by issuing bonds, which would be financed by a surcharge on all insurance, even auto policies, in the state. One estimate puts coastal property value at $2.8 trillion, with Florida assuming $511 billion of that risk through its state insurance fund. Underpricing risk is (or should be) a mortal sin in the insurance industry, and the pool was financially precarious from the beginning. A study of the insurance market conducted by Florida State University concluded, "Florida could be one major storm away from having to take all wind risk."

In the event of a shortfall following a hurricane, Citizens' structure imposes assessments that can extend to all insured properties in the state, even those not at risk from hurricanes. The ultimate backstop is the state's taxpayers. Because its rates are by definition below market (otherwise the private market would absorb all business), the structure of Citizens combined with the demographics of the state implies that poorer Floridians living away from the coast are prepared to subsidize more affluent Floridians who have blithely passed on their hurricane risk to the entire state. One wonders how the portion of the state's taxpayers who live safely away from the coast feel about assuming that risk?

In fact, there was some pushback. The politics that surrounded Citizens became fractious, and remain fractious, but in the oughts and subsequent years, the backstops subsidizing coastal property owners have been sufficiently generous that few moved away. Florida, which had 8 percent of the U.S. coastal population in 1960, according to the U.S. Census, had about 15 percent in 2000 and nearly 16 percent in 2008—again, after the damaging hurricanes of 2004 and 2005. In Florida the main disincentive to moving to the coasts was rising prices, not because of increased insurance but because, with the exception of the collapse during the Great Recession, people were willing to bid up prices to live there.

Behind the scenes, there were increased stirrings in the insurance industry. During the decade, the reinsurance industry did begin to move toward adjusting to more frequent storms—after the 2004 and 2005 hurricanes, reinsurance rates in Florida rose 25 percent or more—but these efforts were largely countered by state programs that moderated the increases passed on to customers. Property owners remained largely oblivious to the economic threat of climate change.

I was involved in one of those behind-the-scenes efforts in 2004 and 2005. In a report entitled *Climate Change Futures: Health, Ecological and Economic Dimensions,* the Harvard Medical School's Center for Health and the Global Environment teamed up with Swiss Re and the United Nations Development Programme to develop scenarios for how impacts from climate change might unfold in several different spheres. I helped edit the report and write the Executive Summary. The report anticipated a lot of the impacts that have become prominent in the seventeen years since it was published. These include the spread of infectious disease, the die-offs of reefs, an increase in nonhurricane windstorms, and even the increase of wildfires in the American West.

Most noteworthy, however, was what might be regarded as one of the report's failures. At the outset, the sponsors hoped to price the economic impacts of rapid climate change. It's one thing to project the economic impact of global warming if it progresses at a stately pace (as the 2006 *Stern Review* tried to do), but it's an entirely different problem if the model for climate involves significant changes and their derivative impacts over just a few years, the new paradigm of climate change that scientists had adopted. In the case of *Climate Change Futures,* the contributors ultimately threw up their hands. The problem was that with rapid climate change a number of nonlinear effects came into play, and while the authors could talk about "discontinuities" in the climate system, about runaway positive feedbacks, and about "stepwise" climate shifts, the unpredictability and myriad possible interactions of these factors made it near impossible to integrate them into a precise scenario of future economic impact.

This should have been a loudly ringing alarm, but like many other such alarms it was ignored outside the climate change community. Still, if the oughts were yet another lost decade in terms of real action on climate change and public concern about the threat, behind the scenes, innovations in finance, technological progress, and a dawning realization in the leadership of developing countries that they too were in the crosshairs of global warming all set the stage for important developments during the second decade of the new millennium. It was not long after the oughts ended that the reality of climate change and the price it was imposing became impossible to ignore.

PART FIVE

2010s: Things Get Real

Fifteen

THE 2010s:
Reality Bites

Widespread public alarm about climate change finally began to appear in the 2010s. Unfortunately, so did some of the negative impacts of climate change that had not been expected until the 2050s. In fact, it was the precocious schedule of climate impacts that finally brought about a public awakening during this time. By mid-decade the reality of climate change could no longer be ignored; it was ruining middle-class lives and costing people a lot of money. That it took more than two and a half decades for Americans to awaken to a hazard that was already happening casts a spotlight on the tragic mismatches of the four clocks. Our inherent blindness to the threat of global warming until it was upon us has fated us to suffer its consequences for decades to come, at best.

Right off the bat, 2010 set a record as the hottest year ever. This might have set some alarms bells ringing, but early in the new decade, much of the coverage of climate change was dominated by news that global warming had "paused," or was on "hiatus," and that the earth had not warmed since 1998. As it turned out, this was not the case but rather an illusion created by statistical prestidigitation. Because it came on the heels of another fake scandal, Climategate, because a new IPCC report made a roundabout reference to a slowdown in its fourth assessment, and mostly because the usual squad of deniers and their enablers in politics and the press were there to promote any contrary information on climate change, the alleged pause got enormous attention.

Major outlets such as the *CBS Evening News*, *The Economist*, Reuters, and, of course, Fox News bought the denier spin, running stories about how the earth had stopped warming and that scientists didn't know why. Chris Mooney, in a rundown of the piling on for *Grist*, cites this insufferably smug bit from a *CBS Evening News* segment: "At the outset of the segment, CBS's Mark Phillips intoned: 'Another inconvenient truth has emerged on the way to the apocalypse. The new United Nations report on climate change is expected to blame man-made greenhouse gases more than ever for global warming. But there's a problem. The global atmosphere hasn't been warming lately.'"

Except it had. The illusion of the pause derived from a few factors, principal among which was that 1998 was an extraordinarily warm year because of a historic El Niño. So, if you arbitrarily took your starting point as 1998, the next six years looked cooler because they weren't as warm as 1998, even though every intervening year was warmer than normal. Another problem was that there were exceptionally few temperature records from the large swath of the globe— the Arctic—that was warming faster than any other region of the planet. Testifying that the earth was heating up steadily during the "pause" was a continued warming of the oceans, which absorbed 90 percent of the excess heat being produced by human greenhouse gas emissions.

For all the talk about the mainstream media being alarmist on global warming, my experience, having watched this story unfold from its outset, is that many of the most respected outlets have been remarkably credulous of claims by deniers, even those claims that fall apart with the slightest scrutiny. Editors, including those without an ideological agenda, love to poke holes in the conventional wisdom, and many don't have sufficient scientific background to see casuistry when it's presented as though it's a serious argument by someone with a PhD. Editors are also competitive, and if the global warming story was collapsing, that would be big news. Consequently, the "pause" had its moment, and for a brief period the message delivered to the public was "Stop worrying about climate change."

This could have teed up still another lost decade, but this time nature intervened. The "pause" percolated in denier circles for several years before it exploded as a mainstream media story as a result of unartful language in the Summary for Policymakers of the IPCC assessment being prepared in 2013. But then every year starting in 2013 recorded average global temperatures in the top ten for records dating back to the 1880s, with many of the years setting a new record, only to have it be broken the very next year. With each passing record year, it became more difficult to sell the notion of a pause in global warming with a straight face. The 2010s turned out to be the hottest decade ever recorded.

The "pause" soon dropped from the headlines, but the denier community has never been one to put aside a good argument just because it's proven false. Even in 2020, denier-friendly blogs were dredging up papers that discussed the hiatus and even "a slight cool-ing" during the 1998–2012 period. Just to be clear, that period had record- or near-record-setting years in 2000, 2005, 2009, and 2010, and every other year was significantly above normal. The only sense in which there was any cooling was that a few very warm years did not set records. In fact, *every* year during that period would have set a new record for warmth in the 1980s, the decade during which the signal of global warming first began to emerge from the background noise.

The 2010s served up plenty more besides heat. In 2012, Hurri-cane Sandy hit the Northeast, overwhelming defenses that had with-stood storms for more than a hundred years. With Sandy, as with so many other storms, it wasn't the wind so much as the storm surge that wreaked havoc. The hurricane hit New York with 90-mile-per-hour winds, moderate by hurricane standards, but it was the largest hurricane ever recorded in terms of extent, and with more than 1,000 miles of ocean to work with, its easterlies, north of its center (Sandy made landfall near Atlantic City), could gather immense amounts of water and push that water toward the coast. With exquisite timing, the storm surge on October 29, 2012, hit New York Harbor at high tide during a full moon. Measured at 14 feet, the surge broke the old record for Battery Park by 4 feet.

Even this, however, wouldn't have overwhelmed New York's sub-ways but for the added push of climate change. Sea level rise resulting from global warming gave the storm surge the boost it needed to make the floods record-breaking. The components of the surge were a 4-foot 6-inch tide and 9 feet 4 inches of storm surge on top of more than 1 foot of sea level rise in New York Harbor (sea level rise in the harbor exceeded average global sea level rise for reasons that will be discussed in the next chapter) since the subways were built. The combination of storm surge and high tide still would have flooded the subways without the higher sea level, but the raised base of sea level meant that the subways would have flooded even with less than a high tide and full moon.

As it was, the storm flooded nine subway stations and four of fourteen subway tunnels (two of which were managed by PATH rather than the MTA). Sandy inflicted $5 billion in damage on the subways, $19 billion on New York City, and more than $74 billion in total costs. A study published in *Nature* in 2021 estimated that sea level associated with climate change increased Sandy's damages by $8 billion. The other ways in which climate change contributed to Sandy's power are more difficult to tease out, but sea level rise was a factor obvious to anyone who could master addition. The engineers who designed New York City's defenses in the nineteenth century to withstand 1-in-100- to 1-in-400-year events could not have anticipated that climate change would bring a rise in sea level and more frequent and intense storms. They didn't know that those events that were vanishingly rare in the nineteenth century might come about every four or five years once global warming began having an impact 125 years later, a nearly hundredfold increase.

I live near the Hudson River at its widest point, about 20 miles north of Manhattan. In the years since the river has been cleaned up (the result of one hundred years of efforts, and one of the great environmental success stories of all time), restaurants and promenades have proliferated along its banks. One of these, Red Hat on the River in Irvington, New York, occupies a former factory building of Lord and Burnham—a company celebrated worldwide for its greenhouses and conservatories—and sits on recovered land a few feet above high-water level (the Hudson is tidal for much of its length). As Sandy's storm surge boiled up the river, the waters rose over the sea wall and continued rising to a level about 6 feet above the floor of the restaurant (the owners put a mark on a door to show where the flood-waters topped out). Given the width of the river at that point—close to 3 miles—that high-water mark represents an awesome amount of ocean pushed upstream.

Lest anyone miss the lesson of Sandy, nearly three years later, Hurricane Joaquin impacted the South Carolina coast—and just stayed and stayed. In this case, it wasn't storm surge so much as rain-fall, 27 inches over four days, that inundated Charleston and other areas. This too was related to climate change, but in a more attenu-ated way, and it took some years for climate scientists to understand the geophysics of the connection.

But the data points were piling up. In 2017, Hurricane Harvey hit Houston and then stayed and stayed and stayed. While it was there, it dumped rain continuously, 60 inches of it on some of the surround-ing towns. The next year, Hurricane Florence hit the Carolinas with a 10-foot storm surge, and then it just stayed and stayed and stayed. Ultimately, it dumped 3 feet of rain on some districts. At one point, Florence was moving so slowly that the hurricane wouldn't have kept pace with a pedestrian out for a brisk stroll. Yet another leisurely walker was Hurricane Sally, which hit land near Gulf Shores, Ala-bama in 2020, ultimately dumping 2.5 feet of rain on the Gulf Coast as it ambled inland at 3 miles an hour. Ida, the first major Atlantic hurricane of 2021, had all the attributes of storms in the climate change era—rapid intensification, very slow movement, major storm surge, and enormous rainfall. After devastating Louisiana, the storm still retained enough punch to paralyze the Northeast, shutting down subways, railways, airports, and highways from Philadelphia to Connecticut.

Forty years ago, these rainfalls would have been dismissed as unimaginable. I remember reading a *New Yorker* article about an epic rainfall in the South. The storm that prompted the article dropped a foot of rain over the course of a few days. These days a single thunderstorm can dump a foot of water in a matter of hours.

Climate change has contributed to the spate of historic downpours in both direct and indirect ways. Warmer air holds more water, but the new and largely unanticipated factor, the factor that has turned downpours into biblical floods, has been a marked slowdown in storms once they make landfall, which leaves them hovering over areas for days at a time.

The key to the slowdown in these tropical storms may lie far to the north in the Arctic in the perturbations of the jet stream that have resulted from global warming. The warming in the Arctic, accelerated by the reduction in sea ice, has reduced the contrast between Arctic air masses and temperate air to the south. This contrast invigorates the jet stream; the stronger the contrast (or temperature gradient), the faster the jet stream. Jennifer Francis argues that as the temperature gradient decreases, atmospheric waves get larger but also slow down, making weather patterns more persistent. In other cases, the jet stream retreats northward, leaving storms stranded with no upper-level winds to steer them on their way.

The decade's series of record-setting rainfalls were the result of a cascade of repercussions related to the warming of the globe. Perhaps even more remarkable, this same set of connections between the Arctic and lower latitudes played a major role in another set of climate-related catastrophes during the decade. It turns out that just as the warming of the Artic contributed to massive floods, it also exacerbated the unending succession of wildfires that afflicted the American West during the decade.

There is no more dramatic evidence that climate change is here than the hellish spate of wildfires in recent years. In California, five of the ten largest wildfires in the state's history occurred in 2020 alone, and all but two have taken place since 2010 (and, as noted, the oughts had their own series of record-setting fires). The largest wildfire in Australian history torched the country in the Australian summer of 2019–2020. Truly massive fires have swept through Siberia and the Russian Far East. By the end of the decade, it seemed that the world was on fire.

Those who would deflect attention from climate change point to increasing numbers of people moving into fire-prone areas (though, of course, this would not apply to the sparsely populated Russian tundra) and to the buildup of dry wood due to misguided fire suppression policies in places like the American West. But these deflections don't deflect. People have been moving into dry parts of the West for decades, and the fire suppression policies date back a hundred years (and, in fact, for more than two decades foresters have been trying to reduce the buildup through controlled burns). What is different is that temperatures have risen, winds have picked up, and protracted and repeated droughts have become more frequent. The extremely dry Santa Ana and Diablo winds, which can flow down from the Sierras with near hurricane strength, heating and desiccating vegetation along the way, are becoming more prevalent in winter, which prolongs the fire season into the normally wet months (this was evident during the massive Thomas Fire in California's Ventura

County, which started on December 4, 2017, and burned well into January).

Teasing out the role of climate change in these fires involves rapidly evolving science in real time. Describing what happened requires setting the context of what might be described as the "standard" weather pattern for the United States in winter.

With the vast Pacific Ocean to the west, the normal weather pattern for the West Coast is for the jet stream to veer northward as it nears North America. That brings warmer air south of the jet stream to California, Oregon, and Washington. Typically, the upper-atmosphere winds then travel over Northern Canada, steered into the continent near southern Alaska by the counterclockwise flow of the Aleutian Low, which strengthens over winter. Once over north-central Canada, the normal-year jet stream then starts south, bringing cooler air to the central and eastern states of the United States. This is why East Coast cities tend to have colder, snowier winters than West Coast cities at the same latitude. This flow gets interrupted by Pacific storms and other factors, but then reestablishes itself.

Things began changing after 1980. The Aleutian Low started to become more erratic in terms of when and even whether it would appear, while the more southerly high-pressure ridge, which usually lasts a matter of days before being disrupted, has been extending northward and becoming so pronounced that it actually blocks Pacific storms from hitting the West Coast by diverting the jet stream storm track much farther north than usual. This causes the jet stream to pick up much colder air in its transit through the Arctic and deliver it to the midcontinent and East, leading to the series of protracted cold spells and "snowmageddons" we've experienced in recent years.

Moreover, the ridge started to persist long beyond the usual life span of such systems. Typically, a high-pressure ridge might last for a matter of days before being disrupted by storms or other events. Beginning in the 2010s, these ridges would form and persist for weeks, even months. Indeed, in 2013, a ridge formed and lasted the entire winter. This was repeated in the following winter, and the winter after that. The phenomenon was so extreme that in December 2013, Daniel Swain, then a graduate student at Stanford, dubbed it "the Ridiculously Resilient Ridge." The name stuck—often shortened to the RRR—and even made it into the scientific literature.

By diverting storms away from the coast, the ridge had other impacts. Protected by the calm air under the high-pressure dome, a blob of warm water surfaced and also persisted. Over seven hundred days it grew to gigantic proportions, ranging between 600,000 and 1 million square miles of ocean (nearly four times the size of Texas) and extending down to 400 feet. Dubbed "the Blob" by research meteorologist Nicholas Bond, this enormous mass of warm water had its own impacts. Apart from enhancing West Coast droughts, the Blob proved to be a mass murderer of marine life up and down the West Coast from California to Alaska. (In the southern hemisphere, an Australia-sized blob has been a persistent presence in the Pacific east of New Zealand, and its attendant high-pressure ridge has been a factor in the decades-long drought afflicting parts of Chile and Argentina.)

During increasingly rare "normal" years, the jet stream will steer Pacific storms into the Gulf of Alaska and environs, stirring up the ocean waters. In turn, the mixing brings nutrient-rich colder waters to the surface to the delight of tiny organisms called krill, which feast on the phytoplankton. When a blob forms, no such stirring takes place, and the entire oceanic food chain suffers. Baleen whales can't find krill; petrels starve; salmon, which eat krill and the fish that eat krill, starve; orcas that depend on salmon as a staple of their diet suffer, as do sea lions and countless other creatures. Not surprisingly, there's been an increase in mass die-offs of gray whales, seals, and various types of birds coincident with the Blob.

One vivid example of this disrupted food chain was recounted to *Audubon* magazine by Kathy Kuletz, a wildlife biologist with the U.S. Fish and Wildlife Service. Short-tailed shearwaters, Pacific Ocean seabirds, typically migrate in April, from their breeding grounds on islands off Australia northward to the Bering Sea, to feast on the explosion of small sea life following the winter mixing of the northern waters. It's a trip of more than 11,000 miles, and the species takes the risk of the journey because their ancestors had prospered from the bounty that awaited them. Because there are only thin pickings over much of their journey (shearwaters will dive more than 200 feet down to find a meal), the birds don't have much in the way of reserves when they arrive up north. When the Blob shuts down mixing, the shearwaters find no food at the end of their epic journey, and they starve by the hundreds of thousands.

Then there are the second-order impacts. The warm Blob waters fostered a huge algal bloom in 2015. The bloom included an alga that contained the neurotoxin domoic acid. The vast bloom shut down crabbing, clamming, and fishing operations from Alaska to Southern California. Another second-order impact involved increased crab-line entanglements and ship collisions as humpback whales moved closer to shore to eat anchovies as krill declined in the open water.

Three years after the first blob broke up, a new blob formed in 2016. Then in 2020 Blob 2.0 (as it was called) arrived and grew to be as big as the first one. Its impact came even as the lingering impacts of the first marine heat wave (a more formal name for blobs) hit ecosystems and fisheries. For instance, Chinook salmon born during the first blob found nothing to eat once hatched. In 2019, three years after hatching, the Chinooks should have begun returning to their birth rivers to spawn. Few did; 2019 saw the lowest Chinook salmon run ever recorded.

The marine heat waves and die-offs are a worldwide phenomenon, a stark reminder that creatures adapted to a particular climate cannot simply move or change their ways when circumstances change (marine heat waves are also responsible for the increasing number of coral die-offs, including the massive one imperiling the Great Barrier Reef). Halfway around the world from the Northeast Pacific, in 2015, twenty thousand endangered saiga antelope died in a matter of days on the Kazakh steppe when temperatures soared to more than 100 degrees Fahrenheit. The cause of death was blood poisoning as the heat and 80 percent humidity caused ordinarily harmless bacteria to migrate from the animals' tonsils to the bloodstream. In Australia in 2014, extreme heat killed forty-five thousand flying foxes (fruit-eating bats) in one day. Not all mass die-offs are caused by climate change, of course; many can be traced to other human impacts such as toxic pollution. The accelerating number of die-offs, documented by a study published in the *Proceedings of the National Academy of Sciences* in 2015, underscores the fact that the fragile balancing acts that sustain most of the world's ecosystems are teetering badly, with climate change increasingly causing the wobbles. These mass mortality events show us that creatures that have adapted over thousands if not tens of thousands of years to particular climatic and oceanic conditions cannot simply adjust when those conditions change abruptly.

Humans are supremely adaptable, but, given our numbers, not all of us can adapt either. Joseph Stalin is reported to have remarked, "If only one man dies of hunger, that is a tragedy. If millions die, that's only statistics." A corollary of that remark for the news business is that if a neighbor dies of hunger, that merits the front page, but if hundreds of thousands die on the other side of the globe, that's a paragraph on page 24. Unfortunately, that cynical rule of thumb applies very well to perceptions of climate change. While climate-related news in the United States has focused on people uprooted by wildfires or flooding, in faraway places such as Iran, large tracts of land have been rendered uninhabitable by extreme heat, forcing farmers to migrate to other parts of their country. Often the migrants end up moving to areas that are also suffering the impacts of climate change. In Pakistan, if it's not the heat, it's the recurrent floods forcing migration, floods caused by an extreme monsoon in 2010, or outburst flooding from mountain lakes pushed beyond their capacity by melting glaciers. In the Pacific, entire island nations, such as Kiribati, are slowly submerging as seas rise.

Migration was much in the news during the 2010s. Europe experienced waves of migrants, many of whom came by boat. They were fleeing war, civil unrest, and persecution in homelands including Iraq, Syria, Afghanistan, Eretria, and other unstable areas. At times, the flood of people overwhelmed Europe's ability to absorb them—in 2015 Greece alone received more than eight hundred thousand migrants and refugees—and the human tide precipitated a rise of anti-immigrant, populist politicians who became destabilizing forces in their own right.

Obscured by the headlines of migrants drowning or being cast adrift by human traffickers was a marked increase in internal migration caused by weather. In 2017, 68.5 million people were forcibly displaced according to the World Bank, and an estimated one-third of these displacements were caused by "sudden onset" weather-related events. More pernicious and even less noticed were migrations caused by weather that was not "sudden onset" but rather an inexorable rise in temperatures and/or drop in precipitation.

In Iran there are different types of heat. In the desert, temperatures soar, but the humidity is quite low, which makes it possible to work at relatively high temperatures. But then there was that day in 2017 when the temperature in Ahvaz in Khuzestan Province, Iran, reached 129 degrees Fahrenheit, the highest temperature ever recorded in the country. With temperatures that high or close to it, a situation that characterized a good deal of Iran, outdoor work is impossible, and crops wither. If the rising temperatures are accompanied by a drop in rainfall (also the situation in much of Iran), making a living as a farmer becomes untenable as well.

The situation closer to the coast of the Persian Gulf is even worse. On July 31, 2015, in the coastal city of Bandar-e Mahshahr, the temperature soared to 115 degrees Fahrenheit, which doesn't sound so bad—until one considers the humidity. On that day, the dewpoint— the temperature at which water vapor condenses—was 90 degrees Fahrenheit. This yields a heat index of 165 degrees. To put this in perspective, the U.S. National Weather Service warns that sunstroke and heat exhaustion become likely with a heat index of 105 degrees Fahrenheit.

Life-threatening temperatures are occurring in unexpected places, including the Far North. In the town of Lytton, northeast of Vancouver in British Columbia, the temperature reached 121.3 degrees Fahrenheit on June 29, 2021, marking the third consecutive day temperatures set an all-time record high for all of Canada. The record was an astonishing 8 degrees Fahrenheit higher than the previous record (most all-time records are broken by mere fractions of a degree), and the blast heat torched areas in Canada and the U.S. Pacific Northwest where fewer people have air-conditioning. This led to a surge of unexpected deaths in metro areas such as Seattle and Vancouver.

Recent studies have suggested that humans cannot survive when a "wet bulb" temperature exceeds a certain limit. These temperatures are calculated by draping a wetted wick over the bulb of a thermometer and measuring the temperature after the water cools the thermometer. The resulting temperature is the lowest that can be achieved by the evaporation of water. This technique is used to calculate humidity, but it's also used to calculate the limits of human heat tolerance. A study led by Colin Raymond and published in *Science Advances* in 2020 held that once wet bulb temperatures reach 95 degrees Fahrenheit, the human body cannot shed heat fast enough through sweating to maintain body temperature, leading to fatal hyperthermia and heatstroke.

The study notes that the deadly European heat wave of 2003, which ultimately killed an estimated seventy thousand people, had wet bulb readings that never exceeded 82.4 degrees Fahrenheit. It also notes that there were no records of wet bulb readings above 91.4 degrees and only a few calculated near that level. The wet bulb temperature for Bandar-e Mahshahr in Iran on that July day in 2015 was 94.3 degrees Fahrenheit, breaking the previous record by 3 degrees Fahrenheit. It was just slightly below the level that would have produced mass fatalities.

In the European heat wave, as well as in many other events, it was the high nighttime temperatures that caused many of the deaths, particularly among the elderly. These temperatures too are rising to unimaginable levels in many parts of the world. In Quriyat, Oman, situated on the Gulf of Oman, there was a period of fifty-one hours in 2018 when the temperature never dropped *below* 107 degrees Fahrenheit.

The increase in reports of such temperatures suggests that even now, some parts of the planet, including areas that have been inhabited since the dawn of civilization, are now becoming uninhabitable. This was long expected; indeed, the Raymond study predicted that temperatures beyond the threshold of human viability would become frequent events later in this century.

Even now, entire continents are experiencing unprecedented heat. A State of the Climate Report from Australia in 2020 noted that during a fifty-eight-year span there were a total of twenty-four days when the average high temperature for the entire continent was above 39 degrees Celsius (102.2 degrees Fahrenheit), less than one day every two years. In 2019 there were thirty-three such days.

We expect the Middle East and Australia to be hot; we don't expect that the Arctic will be hot. Expectations notwithstanding, in recent years the Far North has experienced increasing numbers of heat waves. In the summer of 2019, an extraordinary heat wave affected much of Europe in July and August. Paris temperatures reached 108 degrees Fahrenheit, and many other cities baked in Saharan temperatures. Helsinki, Finland, hit 92 degrees Fahrenheit, the highest temperature recorded in the city since 1844. Then the hot dome moved even farther north, ultimately covering much of Greenland and bathing the ice sheet in temperatures that reached the upper eighties. The ice responded by melting, some 40 billion tons of it in this one event, an amount that measurably raised sea levels around the world. The stunning summer event was just the latest of a series of heat waves hitting all areas of the Far North. In the summer of 2020, the temperature reached 100.4 degrees in the town of Verkhoyansk in the Russian Far East, marking the first time in recorded history anywhere in the world that the temperature had exceeded 100 degrees Fahrenheit above the Arctic Circle. Then, in the summer of 2021, the Greenland Ice Sheet was hit with another extraordinary heat wave, marked by record-setting temperatures and massive runoff. Polar Portal, a Danish monitoring group, estimates that the rate of mass loss of the ice sheet in recent years has been four times the rate in the years preceding 2000.

With more heat has come some truly strange side effects. In the Russian Far North, accelerated melting of the permafrost has caused methane buildups under the land surface, leaving the tundra littered with mysterious domes. One survey in 2017 counted seven thousand such mounds. Many of these are former pingos—massive blocks of ice frozen under the permafrost. As the ice melts, methane is released, replacing ice in the underground cavity. Out of sight of humans, many of these mounds either explode or collapse, creating vast craters on what was once a uniform landscape. Nothing short of a nuclear bomb could create craters as large as some of these explosions/collapses, which put holes in the permafrost as deep as a sixteen-story building. The explosive collapse of one mound in 2013 could be heard 62 miles away. The surreal image of an empty landscape being bombed from below offers a foretaste of the disquieting surprises climate change will bring.

Right now, extreme heat events are taking place in regions that already suffer extreme water and heat stress. Iran gets one-fifth the rainfall of the average nation, and Pakistan ranks fifth on the Global Climate Risk Index for nations vulnerable to the impacts of climate change. The wildfires, droughts, floods, storms, and storm surges of the 2010s offered a proleptic view of the future we face. Given the inertia of the climate system, the longevity of greenhouse gases in the atmosphere, and humanity's unwillingness to accept the threat, much less confront it, it is guaranteed that all these events, whether they be sudden or incremental, will become more and more frequent and intense.

The 2010s were a decade during which the evidence that climate was changing became impossible to ignore. It became clear that these changes were portents of mass die-offs in the animal kingdom; forced migrations for millions of peoples whose homelands were rendered uninhabitable by heat, drought, sea level rise, or other derivatives of global warming; and unanticipated expenses in the richer countries as citizens suffered from recurrent floods, windstorms, and wildfires. Reality, the clock that matters most, was telling the world that climate change is here.

Sixteen

THE CLIMATE PICTURE COMES INTO FOCUS

I n 1998, I published a book called *The Future in Plain Sight.* I knew it was a fool's errand to predict the future and so I chose to examine one simple question: whether the future was likely to be more or less stable than the present. In stable times, innovation and investment flourish, people look outward, cultural identities blur. In unstable times, innovation and investment shrivel, people turn inward and take out "insurance" of various forms, such as strengthened family and community ties. Society as a whole turns inward. The book outlined factors that would bear on future stability, one of which was climate change.

I included scenarios imagined for the year 2050 in order to dramatize how increased instability might transform life. My scenario from 2050 tried to show how forces set in motion by climate change might kill off California's remaining redwoods, trees that had persisted on the West Coast for eight million years. Now, twenty-four years after publication, that very scenario seems to be unfolding.

One culprit is blobs. There have in fact been many bloblike

marine heat waves around the world in recent decades, and a number of factors can cause extreme ocean warming, including an El Niño. Of particular interest, however, is the original Blob in the Northeast Pacific because it likely connects to changes in the Arctic, and those changes have produced a massive cascade of repercussions that are now threatening humans and entire ecosystems. At the tail end of this whip are the redwoods and their relationship with fog.

Redwoods are some of the largest, most durable life forms on earth. Weighing as much as 2,000 tons, they have developed fire- and pest-resistant bark and a root system from which a new redwood will grow even if the tree falls. There are living redwoods whose roots took hold at the dawn of human civilization. The tree's strategy is to grow as tall as possible as rapidly as possible, stealing sunlight and moisture from competitors. The Achilles' heel of that strategy is that the trees are poor pumpers of water, meaning that their upper branches and needles need to supplement what the trees get from their roots. This strategy was fine millions of years ago when the West was wetter, but as the West dried out, the redwoods died out everywhere except the coast, where they could get water from storms during the winter and from fog during the summer. So it turns out that one of nature's most durable creations is utterly dependent on one of nature's most fragile and evanescent atmospheric phenomena.

California's coastal fogs have been a summer phenomenon for millions of years. That's because the fogs are built on two profoundly stable pillars, the coast of a continent and an ocean, as well as one other pillar that proved very reliable until just very recently—trade winds. When things are operating normally—say for most of the past several million years—winds blowing from west to east over the North Pacific drive water toward the coast, causing cold water to well up to replace the surface water. When one current bumps into North America near southern Oregon, it splits in two, with one portion, dubbed the California Current, heading south along the coast, again drawing up colder water from below and cooling the air above it. The colder the air, the less capacity it has to hold water vapor, and when the air temperature drops to the dew point, as it typically does just before sunrise along the California coast in summer, a marine layer will take shape. Then, during the summer day, inland California heats up, and the rising hot air over the interior drags the fog in over the coast, bathing the redwoods in a protective, foggy dragon's breath.

The fogs are a gift of a stable planetary setup of high- and low-pressure systems that extend across the Pacific, driving the current. The frequent reappearances of the Blob—three since 2013—represent an interruption of that setup. When the Blob appears, fogs retreat. This is the tail end of the whip, and it is open to question how many lashes redwoods can withstand.

The handle of the whip is global warming. More than 90 percent of the excess heat created by our overloading the atmosphere with greenhouse gases has ended up in the oceans. There, it has either amplified or set in motion myriad impacts, most of which are invisible to us on land. Let's go through some of the most significant to show how the handle of the whip sets in motion lashes at the other end.

In simplest terms the warming of the oceans has raised the baseline for marine heat waves. Just as sea level rise (itself a product of warming of the oceans) raised the baseline so that storm surges have begun to overwhelm coastal defenses that stood for more than a century, so too has a warmer ocean helped push marine heat waves to extreme levels.

Oceanographers and geophysicists view the ocean and atmosphere as a coupled system, constantly interacting with each other. Climate change has altered these interactions, and some of the most significant changes are occurring in the Arctic. It only accounts for about 3 percent of the earth's surface but, because of its unique composition, the Arctic has a disproportionately great influence on global climate. As noted previously, it has played an outsized role in some of the most extreme climatic events since the end of the last ice age. As the sea ice retreats, allowing the now dark surface of Arctic waters to absorb and release heat, and as the snow season shortens, allowing the same phenomenon to work on the land, the positive feedback of warming begetting more warming has resulted in temperatures in the Arctic increasing at several times the rate of the lower latitudes. This Arctic amplification has been long noted and much studied.

During the 2010s, a number of researchers, including Jennifer Francis, Marilena Oltmanns at the National Oceanographic Centre in the United Kingdom, and many others, published studies connecting the retreat of sea ice and the warming of the Arctic to other anomalous events around the world. One of these was the proliferation of warm blobs in the Northeast Pacific. A number of influences contribute to the formation of warm blobs, but among those the most dramatic has been the warming of the Arctic.

It turns out that apart from causing marine die-offs and aberrant weather, the changes in the Arctic also can interfere with the coastal fogs that have sustained redwoods for millions of years. As noted, the jet stream's vigor is determined by the temperature contrast between the Arctic and the lower latitudes. As the Arctic has warmed, that contrast has diminished, which has slowed the jet stream and dampened its meanders so that it maintains a configuration for a longer period. As we have seen, a slower, more persistently patterned jet stream means that a weak high-pressure system that ordinarily forms in the Northwest United States in the fall and winter can both intensify and remain in place for far longer than it used to.

As this ridge becomes larger, it sets in motion a self-reinforcing cycle. Storms get diverted to the Far North, while the high pressure dampens winds over the Pacific. This lessens ocean water mixing, which allows the Blob to both expand and become warmer. In turn, the warm water bolsters the perpetuation of the high-pressure system. This cycle impacts the California Current and its related fogs.

With less wind to drive the system, the California Current warms and weakens, and with warmer temperatures, the marine layer becomes less reliable, and the West Coast bakes. The redwoods, some of them more than two thousand years old, clinging to their last redoubt on a band of the Oregon and California coast, find themselves struggling for the moisture that sustained them through millions of summers. The tail end of the whip.

As noted earlier, the advent of the Northeast Pacific warm Blob and its accompanying ridge are also implicated in the series of extremely cold and stormy winters in the U.S. Midwest and Northeast in recent years. Just as the RRR and its ilk push the storm track farther north, the distortion of the jet stream has also caused it to roust cold that should have settled in over the Arctic and ship it to the midlatitudes. The phenomenon has been misnamed "polar vortex." The polar vortex is actually a barrier of sorts, formed by the steep temperature gradient that ordinarily bottles cold air in the Arctic and that defines the track of the jet stream. As that temperature gradient decreases, that barrier can break down and Arctic air can spill south, creating the bizarre situation in which the Arctic is bathed in unseasonable warmth while the lower latitudes freeze.

The repercussions of a warming Arctic can become truly confusing. For instance, 3,000 miles east of the Blob in the Pacific Northwest, another blob has been forming in recent years in the North Atlantic. It's also related to the warming of the Arctic, but, paradoxically, this blob consists of intensely cold water and centers southeast of Greenland. Seen from space in satellite imagery, it shows up as a blue blob, one lonely instance of cooling against warming oceans throughout the Northern Hemisphere. Except it's not ocean water, and it poses a potentially catastrophic threat to northern Europe.

One artifact of the warming of the Arctic has been the accelerated melting of the Greenland Ice Sheet. The meltwater from the ice sheet as well as from the many glaciers in eastern Canada flows into the North Atlantic and is collected and pooled by ocean currents. This fresh water is lighter than salt water and tends to stay on the surface. And even though it is meltwater, it is far colder than the ocean water it sits atop. The profoundly disturbing fact is that it shouldn't be there.

Ordinarily, this is near the part of the ocean where the remnants of the Gulf Stream sink, forming what is called "deep water" as part of the planetary system of a river of water within the oceans called the thermohaline circulation. What happens during this part of the circulation is that as the Gulf Stream travels northward it steadily evaporates, releasing heat, and, as it does, it becomes saltier than the surrounding ocean water. More saline water is heavier than less saline water, and at a certain point in the North Atlantic between

Greenland and Iceland it begins to sink. This process is partially a function of a sill on the bottom of the ocean. As the saltier water spills over that sill it "entrains," or drags, the rest of the current behind it. This entraining is a critical part of the heat distribution system of the ocean. The North Atlantic part of the thermohaline system is called the Atlantic Meridional Overturning Circulation, or AMOC.

As Wallace Broecker, Gerald Bond, Willi Dansgaard, and others have reconstructed events, in the past this part of the circulation has been periodically interrupted when the surface of the ocean has been flooded with fresh water from the melting of ice sheets or, in some cases, flotillas of icebergs. When that has happened the thermohaline circulation has shut down, and that in turn has plunged much of the Northern Hemisphere into a deep freeze.

That pool of cold fresh water presents a worrisome sign that something like that may be happening now. Michael Mann and colleagues collected data on sea surface temperatures and, in 2015, published a new index for the AMOC in *Nature Climate Change*. It showed that since 1970 the AMOC has lost considerable vigor. Separately, Stefan Rahmstorf, the German climate modeler, has, with colleagues, published several papers on the slowdown. In a 2018 paper in *Nature*, his team argued that shutdowns in the AMOC have resulted in the most rapid and violent climate shifts in the past 2.6 million years. Estimates are that the current has lost about 15 percent of its normal vigor. The problem for humanity is that no one can confidently estimate at what point this system tips into shutdown or what that would mean for those of us living today. Rahmstorf and colleagues have followed up with a new study published in *Nature* in 2021. It asserts that the Atlantic overturning is the weakest it's been in a millennium, and they estimate that it might further weaken by 25 to 45 percent by the end of the century.

For the moment, this cold blob has contributed to more intense storms in Europe, while the slowdown in the AMOC has contributed to a rise in sea level on the northeast coast of the United States. Sea level seeks uniformity, but there are persistent patterns that distort sea levels in areas around the world. As Michael Mann has explained, when the AMOC is vigorous it creates a downslope that draws water away from the coast. Conversely, when it slows down, that slope decreases so that water piles up by the coast, raising measured sea level.

Thus, the Northern Hemisphere enters the 2020s with dueling blobs, both hot and very cold, with the cold blob the product of heating in the Arctic and itself perhaps the harbinger of a sudden cooling in parts of the North Atlantic. For average Americans, and even for those deeply involved in climate change, these competing forces are hard to reconcile.

The heating of the oceans and the tropics may be pushing storm tracks northward, leaving the Mediterranean regions around the world hotter and drier. Three massive zones of evaporation, transport, and condensation (one can visualize this as a series of belts, each taking up close to 30 degrees of latitude covering the distance from the equator to the North Pole) drive poleward atmospheric circulation away from the equator. As the air and oceans warm, these belts have been shifting poleward on both sides of the equator, leaving a rain shadow in some areas where rain used to fall (e.g., the California coast in the Northern Hemisphere and southern Australia in the South) as clouds drop their moisture nearer the poles.

On the other hand, the slowdown in the AMOC may be pushing tropical rain belts southward. Taken together, what is happening is that the interconnected systems that distribute heat and moisture around the planet are adjusting to the increased energy the warming of the planet represents. It is happening now, and people and ecosystems around the world are feeling the adjustments in freakish and extreme weather events.

The future began knocking on the door in the 1980s, and while the public ignored the knocking, climate scientists noticed that the record-setting years might be a signal that climate theory about greenhouse gases and warming might be climate fact. That knocking became a pounding in the 1990s, and the science community recognized that it meant that climate change was not a future event. Even with a global effort to understand what was happening, the changes occurring outpaced that ability of science to connect the dots. The Northeast Pacific Ocean blob showed up, and oceanographers, geophysicists, and meteorologists scrambled to understand how it came about and what it meant. It wasn't predicted.

For all the surprising weather and geophysical anomalies, there were many that science had predicted that showed up much earlier than expected. Perhaps the most dramatic and consequential examples of such unwelcome visitors from the future have been the signals of distress coming from the West Antarctic Ice Sheet. Events there that were predicted to unfold more than a century from now seem to be happening now.

As noted earlier, in the 1990s, scientists saw that the so-called ice streams that transport ice within the West Antarctic Ice Sheet seemed to be accelerating. WAIS delivers ice to the ocean primarily through the massive Thwaites and Pine Island glaciers. Thwaites is about the size of Florida and Pine Island about two-thirds as big. What surprised scientists was the speed with which the glaciers were delivering ice to the ocean and the changes that were occurring within the ice sheets. This discovery set in motion a global scramble to determine what was going on. Also driving this scramble: none of the models could account for the enormous and rapid shifts in sea level rise evident from the fossil record.

In 2011, Jeremy Bassis and C. C. Walker published a paper in the *Proceedings of the Royal Society A*, offering a mechanism for the accelerating disintegration of glaciers. Stable marine ice sheets such as WAIS typically have ice shelves extending into the seas from the ends of major glaciers. If the ice shelf is in a bay it can act as a cork of sorts, slowing the advance of the masses of ice continually being transported toward the sea. Glaciologists noticed that the breakup of the Larsen B Ice Shelf and the breakup of the ice tongue in front of Greenland's gigantic Jakobshavn Glacier resulted in a dramatic increase in iceberg calving. Bassis and Walker offered an explanation for the increase. They argued that the breakup of such ice shelves exposed ice cliffs in the main body of the glacier and that the structural properties of the ice could not sustain a cliff more than about 300 feet high. Once exposed, an ice cliff taller than that will fracture and collapse from its own weight, leading to a flood of icebergs.

Since this argument was put forward, a number of scientists have refined and developed the theory, linking it to other aspects of ice transport toward the sea. While uncertainties remain, the picture that has emerged during the 2010s is that as the ocean and air warm, melting and runoff deliver water into and through the ice shelves, leading to a complicated series of interactions that ultimately thin and weaken the shelves from below. Once they disintegrate, the back pressure on the glaciers is relieved and they begin to move more rapidly toward the sea.

The disappearance of the ice shelves leaves an ice cliff exposed at the terminus of the glacier, and at this point a self-reinforcing process can lead to runaway acceleration of the glacier. In the case of the Thwaites and Pine Island glaciers, the more ice that breaks off, the higher the remaining ice cliff (because of the slope of the land underneath the glacier). The higher the cliff, the more ice will hive off.

Thwaites alone now contributes about 4 percent of annual sea level rise. Its acceleration has already forced reappraisals of future sea level rise that themselves are just a few years old. For instance, a study led by glaciologist W. Tad Pfeffer published in *Science* in 2008 estimated that the Antarctic might contribute 60 centimeters of sea level rise by 2100. By 2013, that estimate was raised by 50 percent to 90 centimeters. Expect it to be raised again.

The push to understand Thwaites is yet another example of what is becoming an all-too-familiar phenomenon. In the space of a few decades, what was isolated speculation about something long in the future became a present-day threat and an urgent matter for science in real time. So urgent that the United States and the United Kingdom have jointly set up a mission to study Thwaites. The International Thwaites Glacier Collaboration involves sixty scientists who over the next five years will try to understand the ways in which this giant unstable glacier might contribute to rapid sea level rise.

Over the past three decades, the vast global network of scientists and scientific institutions has mobilized to understand all the ins and outs of the global climate system and its vulnerabilities. Writing for *RealClimate*, Stefan Rahmstorf pointed out that in 2018 there were roughly twenty thousand peer-reviewed papers published on climate change. That works out to fifty-five papers a day, more than two every hour for every day of the year, meaning that studies are accumulating far more rapidly than any one human could read them.

The picture emerging from these thousands of studies is of a climate system quite different from the one we thought we had, and also of a climate threat far more dire and more imminent than was anticipated by the IPCC's first assessment in 1990. Then, the Greenland Ice Sheet seemed stable, and the conventional wisdom was that the Antarctic ice sheets might well grow as the climate warmed. In 1990, sea level was rising at about 1.2 millimeters per year, not far off the same pace it had been rising for more than a century. At that time, the paradigm for climate change was that it would arrive at a stately, incremental pace. It was believed that the permafrost would remain stable for the next hundred years.

Then, as the scientific world mobilized, the news that past climate shifts were violent and extremely rapid was confirmed. In 1996, the Greenland Ice Sheet began losing mass, a process that has accelerated dramatically as the years have passed and is now more than seven times the rate it was at the end of the 1990s, according to the National Oceanic and Atmospheric Administration (NOAA). WAIS was discovered to be losing mass not long after that, and more recently, evidence has surfaced that the East Antarctic Ice Sheet, the mother of all ice sheets, is losing mass as well. By the mid-1990s it became obvious that permafrost was melting throughout the Arctic, and by the 2010s it became conventional wisdom that most of the world's top layer of permafrost would disappear by 2100. By the middle of the 1990s, the rate of sea level rise had more than doubled; by the early part of the new millennium, the rate had tripled; and now it is close to quadruple the rate of rise in 1990.

Indeed, in many cases the worst-case scenarios of the early 1990s have become the conventional wisdom of today. The current intermediate case for sea level rise in 2100 is roughly the same as the high case in the first IPCC assessment. NOAA's worst case of 2.5 meters is more than four times as high as the high case for the IPCC of 2007. Even these numbers are not the worst case.

During the 2010s, the IPCC made big strides in catching up to the science, if not reality. As noted earlier, the reports themselves are supposed to represent the state-of-the-art science, but for most of their history, most notably with the fourth assessment in 2007, the Summaries for Policymakers have been far more conservative than the chapters, and some of the chapters lagged the actual state of knowledge. Perhaps because of the very strong reaction to the fourth assessment's lowball estimates of future sea level rise, subsequent reports have been less equivocal about the severity of the threat the world faces. In October 2018, the IPCC issued a special report, *Global Warming of 1.5°C*, which focused on how hard it would be to keep warming to that level, how harsh those consequences might be for humanity and nature, and how much more harsh life would be should warming continue to 2 degrees Celsius. While written in the maddening, plodding prose of a document subject to a thousand editors, this time its message was loud and clear: you don't want to go there. Then, in August 2021, the IPCC began releasing draft elements of its Sixth Assessment Report, the message of which UN Secretary-General António Guterres described as a "code red for humanity."

Scientific knowledge continues to advance, but there won't be a straight-line path to perfect understanding. Accounting for all the variables that go into weather, much less climate, far exceeds the capacity of even the most advanced supercomputers. A quip that circulated among climate scientists about their work went "It's not rocket science; it's much harder than that." Even as climate scientists discover new connections among the various gears of climate and new geophysical processes that underlie the dynamics of the ice sheets, the climate offers new anomalies and conundrums. Science has mobilized, but its clock will never catch up to reality. The best we can hope for is that science doesn't lose ground as climate continues to change.

Seventeen

THE 2010s:
The Public Realizes That
Something Is Wrong

In recent years, opinion experts plumbing public attitudes toward climate change have argued that after decades of relative indifference, climate change has become a matter of urgent concern for a majority of Americans. But how urgent? The answer, unfortunately, is not that much. Consider the last debate between President Trump and Joe Biden that took place two weeks before the 2020 election.

The good news for those concerned with climate change: the debate moderator, Kristen Welker, put the issue to the candidates. The bad news: the Biden campaign's reaction to one comment in the debate suggests that his advisers believe that Americans are still not ready to deal with the issue. Biden made a remark about transitioning away from oil, not a particularly radical suggestion. Trump gleefully jumped on the remark, saying, "He's going to destroy the oil industry. . . . Will you remember that, Pennsylvania?" Biden could have doubled down and said, "Climate change is already costing Americans hundreds of billions of dollars. We cannot deal with the threat without moving away from fossil fuels." Instead, his campaign panicked.

Biden surrogates went into furious spin mode—stressing that Biden was talking about a *slooow* transition and that oil and fracking will be with us for a long time. The notion that it is political suicide to talk about reducing our reliance on fossil fuels might have made sense thirty-four years ago, when the recognition of global warming first burst upon the scene. But now? The debate took place even as the latest and worst of a series of hellish fires torched the West.

No party nominee for president before Biden made climate change a major theme of his or her campaign. There are many reasons why global warming has failed to gain traction with the public, including a well-funded campaign of obfuscation by deniers, competition for mind space from other pressing issues such as terrorism or the economy, and the very nature of the issue itself. The threat derives from hard-to-explain photochemical processes in the atmosphere, and sometimes it seems that every human activity contributes to a warming globe.

Given these headwinds, only an inspired leader might get the public to vote for action. Many have tried, but none has found the right message. Billionaire Tom Steyer made climate change a centerpiece of his 2020 campaign and gained a couple of delegates after spending $191 million just on advertising. Jay Inslee, the governor of Washington, also put climate front and center and also came up empty.

If the public was sufficiently educated about climate change, the scandal of the debate might have been Trump's bizarre remarks on the subject. Everything he said was either nonsensical (that Biden was promoting "little, tiny, small windows"), wrong (wind energy is "extremely expensive"), or both. Instead, much of the follow-up coverage was about how much damage Biden might have done to himself through his awkward comment about oil.

While the focus for much of the press was whether Biden's "gaffe" on transitioning away from oil would hurt him in Pennsylvania and Texas, the far bigger problem for humanity is that more than two generations after global warming became an issue, a candidate for president still has to treat it like a hand grenade with the pin out.

Of course, the United States and the world have to transition away from fossil fuels! Moreover, that transition is already underway. In the years before COVID-19 hit, the global economy grew by roughly 3 percent a year, while carbon dioxide emissions grew by only 1 percent a year. Some of this decarbonization comes from the switch to renewables and some because of increases in efficiency, but if the world is going to head off the most disastrous impacts of climate change, the trend has to accelerate because the global economy is still adding greenhouse gases to an overburdened atmosphere.

Only the federal government can enact the carrots and sticks (and negotiate with other nations) to speed this transition. That's not going to happen if an incoming administration waffles on the issue. Most disheartening about the Biden campaign's backpedaling is the implication that Biden's advisers still fear that there's more to be lost by bold statements on the need to move away from oil than there is to be gained.

Alas, the wisdom of this was born out in the general election. After four years of dismissing climate change as a hoax, installing climate deniers in key agencies involved with studying and regulating climate change, countermanding or defanging regulations related to reducing CO_2, promoting coal and other fossil fuels, and innumerable other efforts to make sure that the United States could increase its releases of greenhouse gases, Trump received eleven million more votes in 2020 than he did in 2016.

This surge of turnout for a climate change denier occurred just a year after Greta Thunberg, a Swedish schoolgirl, addressed the United Nations and became an international sensation. She and others spawned a global uprising of the young, who expressed their outrage at the older generation whose heedlessness about climate change put their future at risk. The feelings expressed are profound and real; psychiatrists call it "climate grief," and it has caused great numbers of young people to assert that they are not going to have children for fear of the world in which they would have to survive.

Clearly, something powerful was causing voters to ignore these public expressions of climate dread.

If the top five issues for the voting public listed by *FiveThirty-Eight*'s opinion survey actually determined voting patterns, the results of the election would have been very different. Certainly, Biden got the votes from people who were deeply concerned about climate change, COVID-19, the threats to Obamacare, and racial inequality, and a good deal of the votes from citizens worried about Trump's handling of the economy. But those numbers were not nearly as large as the ones that showed up on surveys. Trump voters seemed to be worried or moved by factors not captured by those polls.

Cognitive scientist George Lakoff has written about "deep framing," a phenomenon through which messaging attaches political parties or particular people to archetypal values or repulsions and gives such emotional power to these associations that appeals to reason don't have a chance. If the messenger is illegitimate, then the message received is not the message intended. When candidate Biden talked about climate change as an existential crisis, a good portion of viewers thought, "Oh great, here's another liberal know-it-all telling me what to worry about."

The power of such framing is almost impossible to overestimate. During the COVID pandemic, Trump supporters who bought his spin that the disease was like the flu embraced that belief so fervently that some coronavirus patients refused to believe the disease was real even in the moments before they died of it. Jodi Doering, a South Dakota ER nurse, told CNN that some patients would refuse to call family or friends because they thought they were going to be fine right up to the end. Others' last words would be, "This can't be happening. It's not real." If someone can embrace an alternative explanation so fiercely that they refuse to abandon it even as they are dying of what they deny, then it's easy to understand how those who think climate change is a liberal fantasy can ignore decades of evidence to the contrary.

Those trying to communicate the scale of the climate threat know this, of course. They've seen decades of different attempts to invest the global warming message with some emotional oomph fail to move the broader public. To try to communicate more effectively, delegations of scientists and writers have met with editorial boards (I've been part of some of these efforts), numerous universities have established programs to better translate the complexities of the science into a form that the public might understand, high-powered marketing and advertising wizards have volunteered their time to better shape global warming messaging, and corporate leaders and major financiers have put serious money into attempts to galvanize the public.

These are not trivial exercises. Hedge fund titan Jeremy Grantham dipped into his personal fortune to fund a blizzard of initiatives to try to bridge the gap between the urgency of the problem and the relative apathy of the public. Yale sponsors a climate film festival and has several other programs attempting to communicate climate change science and news. Columbia University also has several programs related to bringing climate science to the public in digestible form, as well as a Climate Decision Forum, an interdisciplinary effort with one sector devoted to studying the social scientific issues involved in communicating the climate threat. I've participated in climate change messaging conferences that brought in linguists, psychologists, and experts on semiotics. Smaller universities such as Appalachian State University in Boone, North Carolina, have put together programs to engage local communities on the issue.

Big foundations such as Hewlett, the Bullitt Foundation, and the Rockefeller Philanthropy Advisors have poured money into web presences such as *Climate Nexus*, Climate Central, and *Inside Climate News*, all of which try to crack the code of how to galvanize the public about the scale of the danger and report on possible pathways to reduce its potency.

Major movie stars such as Leonardo DiCaprio have picked up the baton from Al Gore and participated in documentaries about global warming. Author-turned-activist Bill McKibben has organized thousands of demonstrations around the world. CliFi—shorthand for "climate science fiction"—has emerged as a genre, and global warming is showing up as a regular leitmotif of feature films such as *Beasts of the Southern Wild* and *Snowpiercer*. James Cameron, the director of *Titanic*, assembled Hollywood A-listers including

Matt Damon, Jessica Alba, and Harrison Ford to host episodes for a Showtime series on climate change.

Both in the United States and globally, the scale of the effort has been massive, expertly targeted, and social media savvy. And almost everybody has gotten better at it. Remembering previous efforts to sell the public on the danger of global warming, Bud Ward, the founding editor of *Yale Climate Connections*, remarked wryly, "If you wanted to tell the public that climate change was a faraway threat long in the future, there was probably no better symbol for the threat than the polar bear."

Demonstrably, fine-tuning the outreach efforts has not been enough. As the 2020 election shows, this colossal campaign to communicate the problem has only worked with those ready to receive the message. The problem is not information; we're drowning in new examples of the disruptions and distortions resulting from our changing climate. The problem is a frailty in the American psyche that makes very large numbers of people more receptive to truly lunatic conspiracy theories (e.g., the QAnon belief that Democratic elites are running an international cannibalistic pedophile ring, an insanity believed by half of Trump supporters according to a poll conducted by YouGov just before the election) than to the tsunami of reporting on global warming by the mainstream media.

The New York Times collaborated with Siena College to poll voters on attitudes toward climate in the battleground states of the 2020 election. They asked voters in Florida, "How worried are you that rising sea levels from global warming will have a significant impact on your life?" The vast majority of Biden voters were either very worried (40 percent) or somewhat worried (38 percent). When it came to

Trump supporters (Trump won the state) only 3 percent were very worried and 19 percent somewhat worried. The vast difference in attitudes toward something that already affects life in the state suggests that many Floridian neighbors filtered and interpreted the evidence of their own eyes in radically different ways depending on their party affiliation.

A question put to Arizona voters about whether they were worried by rising temperatures produced similar differences. About 90 percent of Biden voters were very or somewhat worried. Only 22 percent of Trump supporters voiced concern.

Anthony Leiserowitz and Jon Krosnick, a Stanford University-based political scientist who has decades of experience studying opinion on climate change, rightly note that the "issue public" for climate change dramatically expanded in numbers during the latter parts of the 2010s. Despite this uptick, the issue still lacks the political potency to drive a national election. Will the issue public continue to expand sufficiently to drive action on climate now that Biden has replaced Trump as president? After all, smoking never drove a presidential election as an issue, and yet the public turned away from cigarettes. Seat belts never drove a presidential election, and yet the public now wears them without complaint. And recycling has never been a major issue in a presidential campaign, yet almost everyone now separates plastics, metals, and paper from garbage.

Even if it didn't arouse sufficient passion to break through the partisan filters of Trump supporters, concern about the threat of climate change has reached a threshold where Biden will have a receptive public when he reverses the myriad pro-greenhouse-gas actions of the Trump administration. One of his very first acts was to have the United States rejoin the Paris accord on climate change; he has begun reversing the scores of regulatory changes through which the Trump administration attempted to promote the unfettered use of coal and other fossil fuels (as well as many toxic compounds); and we can expect that Biden will promote new initiatives to hasten the transition away from fossil fuels.

Unfortunately, because of the nature of carbon dioxide, we cannot simply pick up where we left off when Trump took office. Just as retrograde action on climate change during the George W. Bush administration narrowed the window for subsequent administrations to deal with climate change, so did the Trump administration's undermining of climate initiatives further raise the burden of CO_2 for Biden.

A story by Christopher Flavelle in *The New York Times* reported that the Trump administration actively tried to sabotage the administration's own report on the outlook for climate change. It tried to soften language and introduce qualifications and then dumped the report on the public the day after Thanksgiving in the hope that it wouldn't be noticed. Testament to the dedication of the scientists who put out the report is that all the various attempts to derail or defang it largely failed. Flavelle quotes Thomas Armstrong, who led the previous administration's research efforts on climate change: "Thank God they didn't know how to run a government."

Still, the two Bush administrations and one Trump administration since 1988 have done a staggering amount of damage. For half of the thirty-four years since climate change became a mainstream issue, the executive branch of the largest economy on the planet has been trying to delay, if not outright sabotage, action on global warming. To those sixteen years, you can add the eight years of Reagan administrations, during which climate change was ignored. For much of that time, the United States found willing accomplices in the leadership of

Australia, Canada, and other supposedly enlightened nations, as well as the support of the petro-states. Yes, China and India bear much responsibility for the path of energy development they chose, but criticism from the United States loses force when our elected leaders have derided the threat for half the climate change era.

Still, the damage might have been far greater. Despite the Trump administration's active promotion of coal and feverish deregulation, emissions remained virtually flat for the first three years of his tenure (as wind and solar continued to take off and electric power companies switched from coal to natural gas) and then actually dropped in 2020 as the COVID-19 epidemic greatly slowed economic activity.

The Rhodium Group estimated the pandemic cut U.S. carbon dioxide emissions by 20 percent (globally, estimates for a COVID-related reduction range around a 5.8 percent reduction for 2020 according to the International Energy Agency). Thus, even as the United States pulled out of the Paris climate accords on November 4, 2020, the dramatic decrease in GHG emissions gave the United States the opportunity to meet the targeted reductions for 2025 that they had just abandoned. Ironically, Donald Trump's bungling of the U.S. response to the pandemic proved more consequential for climate than all of his pro-fossil-fuel actions, almost all of which would be undone by the incoming Biden administration. As Dana Nuccitelli wrote in *Yale Climate Connections*, even a 10 percent drop in emissions (the original estimate for U.S. COVID-related GHG emissions reduction) meant that the Biden administration would have to decrease emissions by 2.8 percent a year from 2021 to 2025 to meet the Paris targets. Not an easy feat, but certainly achievable.

This would be quite good news but for one small detail: the Paris accord's targeted reductions would only slow, not stop, the world's march toward global warming greater than 2 degrees Celsius, with all its unknowns, dangerous thresholds, and tipping points. Various studies have estimated that if the targets laid out in the original agreements were achieved, temperatures would still rise by between 2.7 and 3.7 degrees Celsius. It is not alarmist to say that a temperature rise of 3.7 degrees Celsius would put several billion people at risk of starvation and endanger or collapse virtually every major ecosystem on the planet. And the 3.7 degrees Celsius figure lends a patina of specificity in predicting the relationship between increased GHGs and a degree of warming that is entirely unwarranted. No one can say with confidence at what point a warming Arctic might trigger an unstoppable feedback loop in which the melting permafrost releases methane and CO_2 that produces more warming that releases more greenhouse gases.

This is one Achilles' heel of the Paris accord, regarded by many as the signal achievement on international action on global warming during the 2010s. Another is that its targets are goals, not enforceable requirements.

These compromises were likely necessary in order to get the accords to be adopted by all 196 nations that are signatory to the United Nations Framework Convention on Climate Change (UNFCCC). By itself, this was significant, as it is the only United Nations agreement that was signed by every nation on earth—until the United States temporarily backed out. And, despite the compromises, the agreement represents progress toward global action. It establishes monitoring mechanisms and allows flexibility in achieving reductions. The very nature of the agreement creates peer-group pressures toward compliance, and the world has already seen the fruits of this pressure as numerous nations have announced goals that exceed the agreement's expectations. Nations that have set a target for net-zero emissions include the entire European Union, China, Canada, the United Kingdom, Japan, South Korea, and, so far, fourteen other nations that taken together account for a majority of the world's GHG emissions. The list should include California. It's not a nation, of course, but it is the fifth-largest economy in the world, and its target is to be carbon neutral by 2045.

Most important, the agreement can be strengthened. Voluntary targets can be lowered, and enforcement can become mandatory, either inside the agreement or through climate tariffs. Many countries in the European Union are already lobbying for tariffs on nations that fail to try to limit carbon emissions. Leaders from France and Spain have aggressively lobbied for such tariffs, and France has already threatened to impose tariffs on U.S. goods in response to the pullout from the Paris accord. John Kerry, who led U.S. negotiators on the Paris accord during the Obama administration, told *Politico*, "It's not a question of whether it's going to happen; it's going to happen." As climate impacts worsen, it's likely that nations will be grateful for the Paris accord for providing a framework for negotiating such tariffs as an alternative to multiple trade conflicts popping up around the world.

The pandemic dominated the year 2020. Were it not for COVID, the main story would probably have been climate change. Climate was in the news first because of the record fires and then because of record hurricanes, with thirty named storms in 2020. For only the second time, the National Hurricane Center ran through the alphabet and had to resort to Greek letters for names. A capstone to the extraordinary year was Hurricane Iota, which formed in November, making that the first time two hurricanes had formed in the Atlantic that late in the year. Iota represented the latest date an Atlantic hurricane ever reached category 5 intensity.

One final important development of the 2020s has been that the mainstream press finally seems to understand that every climate-change-related story doesn't need to include a quote from a climate denier—not that the deniers have laid down their weapons. When Texas suffered a historic and protracted deep freeze in February 2021, a series of blackouts forced millions of people to cope with single-digit temperatures without heat and water. The governor, Greg Abbott, and Congressman Dan Crenshaw immediately singled out frozen wind turbines as the culprit, while Fox News blamed the turbines and green policies no fewer than 128 times over the next few days (as tabulated by Media Matters for America). *The Wall Street Journal* also weighed in, warning that such blackouts were the future if Americans adopted green policies. The accusations were flat-out lies—top executives from the state's power grid immediately pushed back, noting that wind power outages accounted for only 13 percent of the power losses, and that the overwhelming preponderance of the outages came from problems related to fossil fuel facilities and pipelines. Most of the mainstream press picked up on this, nipping this meme in the bud before it got traction. Even more heartening, many in the mainstream media noted that the characteristics of this freeze—its protracted nature and the breakdown in the polar vortex that allowed the cold to spill south—bore hallmarks of similar episodes in recent years that were attributed to shifts in the Arctic related to climate change.

I'm continually amazed, though, by the knee-jerk tendency of conservatives to jump on any opening to demonize alternative energy. Texas politicians such as Abbott and Crenshaw are flunkies for the state's massive oil and gas interests, but less understandable are the positions of Fox and *The Wall Street Journal*. After all, a wind turbine is just a piece of electrical generating equipment; it's not Democrat or Republican. It either works or it doesn't, and in Texas, wind has worked very well. Most likely, this is another case of the messenger outweighing the message; it's not the technology that arouses ire but those perceived as championing alternative energy. Given the misery they have suffered, it will be interesting to see how long Texas voters continue to buy into the energy policies the GOP has been selling them.

If COVID often displaced climate change in the news, it also propelled the climate change story in unexpected ways. Apart from temporarily reducing GHG emissions, the pandemic's impact on economic activity gave citizens of the world's most polluted megacities a glimpse of what life could be like. In New Delhi (where air pollution shortens life expectancy by an astonishing twelve years), many millions of people could see the Himalayas for the first time in their lives. China saw similar temporary pollution reductions in many of its megacities.

A taste of clear air gave hundreds of millions of emerging-nation city dwellers a glimpse of the possible, and with renewables now competitive with all fossil fuels, those who wanted to preserve that clarity had alternatives to point to as sources of power as economies rebound. Indeed, Amory Lovins believes that COVID might be the stake in the heart of the financing of future fossil fuel projects precisely because alternatives are ready to pick up the slack. Says Lovins, "We saw peak coal in 2013, peak auto sales in 2017, peak fossil fuel power generation in 2018, peak fossil fuel use in 2019. Then COVID crashed the global economy. When it rebounds, renewables will provide the power. The cost of capital will rise for fossil fuel projects, and the fossil fuel lobby will lose political clout. We're going to see a lot of stranded fossil fuel assets."

If Lovins is correct, the ill wind of COVID may have blown some good.

Eighteen

THE 2010s:
Business and Finance
Awaken to the Threat

For most of the climate change era, the business and finance community has been the biggest impediment to action. With few exceptions, the moneyed interests saw actions to prevent climate change as saddling businesses with more costs, more red tape, more penalties, and depressed profit margins. The more enlightened leaders gave lip service to the threat, but the fossil fuel–related interests dominated the corporate/financial narrative on climate change. Then, in the 2010s, the business and financial communities started to catch up, and very rapidly.

The change is still ongoing as of this writing, but the momentum is undeniable, and the evidence is everywhere. While in the 2000s, coal production in the United States increased during the fossil fuel–friendly administration of George W. Bush, in the late 2010s, even attempts to *mandate* coal use by the Trump administration could not stop the decline of the fuel. Nor could Trump's hostility to renewables stop their rise. In 2020, renewables accounted for about 20 percent of power generation in the United States and for the first time equaled the contributions of coal and nuclear.

The financial markets that direct the flow of capital weighed the prospects of fossil fuel companies against renewables and decided where the future lay. Coal plants became unfinanceable, and scores of oil companies went bankrupt by the end of the decade. In October 2020, NextEra, a power company and the world's largest public company focused on solar and wind, had a larger market capitalization than Exxon, once the largest company on the planet. To complete the indignity, Exxon was kicked out of the Dow Jones Industrial Average. Goldman Sachs projected in late 2020 that by the end of 2021, investment in renewables would surpass investment in oil and gas drilling for the first time ever.

In January 2021, Tesla had a market capitalization of $700 billion, which made it more valuable than Volkswagen (which includes Audi, Bentley, and Porsche, among others), BMW, Daimler, Toyota, Honda, General Motors, and Ford taken together, despite the fact that these companies sold about one hundred times more cars than Tesla did in 2019. The Tesla valuation was absurd, but it was also solid evidence of where investors thought the future of autos lay. Nikola, an all-electric truck start-up, went public and at one point sported a $10 billion market cap without having made a single vehicle.

Now the major car companies are all jumping on the electric vehicle bandwagon. In 2020, Volkswagen announced it would spend $86 billion on EVs over the next five years. Porsche, Audi, and Mercedes all announced new EVs as well. At a conference sponsored by Barclays Capital in November 2020 (as reported by *The New York Times*), Mary Barra, CEO of GM, said during her remarks, "Climate change is real and we want to be part of the solution," going on to say that GM wanted to lead in the production of electric vehicles. Then in early 2020, GM made the dramatic announcement that its vehicle fleet would be all-electric by 2035. (While such commitments from a major carmaker are welcome and significant, it should be noted that GM began experimenting with electric vehicles—the EV1—twelve years before the first Tesla came onto the market. GM discontinued the EV1 in 2002, a year before Tesla was founded and six years before the first Tesla was sold.)

Also, it should not be forgotten that before its conversion to electric vehicles, GM spent more than a decade fighting action on climate change. It was a founding member of the Global Climate Coalition, the lobbying group formed in 1989 to delay action on global warming, and only withdrew from the organization in 2000, after membership became an embarrassment. Even as it announced its commitment to EVs in 2020, GM supported the Trump administration's efforts to roll back California's strict fuel economy standards and only dropped their support once Biden was elected.

Indeed, at the time, GM's early, feeble efforts to develop electric vehicles seemed more designed to prove that EVs weren't practical and had little market appeal. Its conversion to the cause of climate change may have more to do with the fact that it has been humiliated in the marketplace by Tesla, which investors viewed at the end of 2020 as ten times more valuable than GM.

The new momentum of climate action has made for some strange bedfellows. Southern Company, another founding member of the Global Climate Coalition and former funder of climate denialism, joined forces in 2020 with Tesla, Uber, Duke Energy, and other heavyweights to form ZETA, the Zero Emission Transportation Association, to lobby for tax and other incentives to support EVs. Southern hasn't cut its ties to coal or coal lobbying. As of this writing it remains a member of the American Coalition for Clean Coal Electricity and several other groups pushing back against the global warming consensus (one of the coal groups describes CO_2 as "plant food"). Indeed, even as the company pushes EVs through ZETA, it also (through a subsidiary) funds the American Legislative Exchange Council (ALEC), a special-interest group that supported the Trump administration's rollbacks of every previous initiative related to climate change.

Expect such internal contradictions to become increasingly common as industry transitions from seeing climate action as a threat to seeing it as an opportunity. While strategic groups within a company might find opportunities in the trends toward renewables, those tied to legacy assets will continue to try to protect their options. Thus, while the Southern Company brags on its website about its commitment to "carbon-free and carbon-neutral energy sources" and cites that 15 percent of its electricity now comes from renewables, through its support of ALEC, it lobbies against mandates for renewables, even though the company's own goals are not much different from the Californian mandate, one of the most aggressive in the country.

Money flows can create a self-fulfilling prophecy. When investors flock in, companies can spend on R & D and innovate and thereby extend their advantage. As they grow, they hire workers and increase their spending, which, in turn, gets the attention of local, state, and federal politicians, increasing the political clout of the alternative energy lobby. Renewables accounted for virtually all job growth in the electric power sector in recent years.

Examples of this feedback loop abound. A New York–based fuel cell company named Plug Power builds hydrogen power systems for vehicles. It scrambled for money during most of its first eighteen years of life as a public company. From 2010 onward, its share price never got out of single digits. The stock started 2020 at $3.23 a share. By mid-January 2021 it reached $70 a share, giving the company a $31 billion market cap, up more than 2,000 percent in little more than a year (the share price has since retreated, but, as of this writing, remains close to ten times what it was at the beginning of 2020). On the way up, the company filed for a secondary offering to bolster its liquidity. With part of this windfall (nearly $900 million), Plug plans to build what it calls the "Gigafactory" to greatly expand its ability to build fuel cell stacks, and the prospect of new jobs in a struggling part of the state had both Senator Chuck Schumer and New York governor Andrew Cuomo offering incentives and touting the advantages of building the factory in New York.

If money is flowing into one sector, it's also usually flowing out of another. James Murray, writing for *NS Energy*, noted that fifty coal companies have filed for bankruptcy in recent years. This was predictable, but even gold-plated fossil fuel companies have suffered. Saudi Aramco, once the world's most profitable company, struggled to find investors for its IPO in 2019. It had to threaten and bribe investors to get the IPO done, and now it is going deep into debt to pay the $15 billion quarterly dividend it was forced to offer in order to entice reluctant investors. In 2020, Aramco started selling interests in subsidiaries as its profits cratered during the COVID pandemic. Chevron, an oil company with one of the strongest balance sheets of any company in the world, saw its stock pummeled in 2020, even though the oil giant had consistently raised its dividend for thirty years.

The renewable boom has similarities with the dot-com bubble of 2000, which, of course, went bust. In the aftermath of the first tech bubble, however, the solid companies survived and for the past several years have been the dominant forces in the equity markets. Google wasn't even a public company in 2000 when the bubble burst. Something similar may happen with renewables as it is clear that some valuations have become irrational. Unlike the first explosion of the dot-com bubble, however, a lot of alternative energy and EV companies are either profitable or well along the path to profitability.

The money flowing into renewables is consequential, as is the money flowing away from fossil fuels. There have been several false starts for renewables before—in the 1970s, when the Carter administration offered tax credits to spur energy independence; in the 1980s; and smaller blips since. For several reasons the current boom will likely have staying power: because many renewables can compete head-to-head with fossil fuels even without subsidies; because imaginative pricing and financing innovations solve the problem of upfront capital costs for both users and suppliers alike; because technological innovations and advances in battery technology are well on their way to solving the problem of storing renewable energy when the sun sets and the wind dies; and, most important, because the economic damage already being caused by climate change provides a constant reminder that the world must reduce its reliance on fossil fuels.

If the business and finance community focused more on the costs of dealing with climate change for the first decades of the climate change era, that same community during the 2010s shifted to focusing on the economic threat of climate change itself. Some of the first shots were fired early in the decade. In 2014, Illinois Farmers, an insurer affiliated with the European giant Zurich Insurance Group, filed nine class action suits against municipalities in the Chicago area for losses the company sustained when extreme storms caused sanitation systems to back up, showering the interiors of hundreds of homes with sewage. The suit explicitly said that officials in these towns were aware that climate change would bring more extreme weather but failed to take any action to adapt to the new reality. Ultimately, Illinois Farmers withdrew the suits, noting that they had achieved their objective.

The suit served notice that insurers were not going to passively accept risks passed on to them by municipalities. It also underscored one of the profoundly difficult issues of pricing the risk of climate change. Ostensibly, Illinois Farmers was acting on behalf of the homeowner victims, but had they pursued and won the suit, the increased costs incurred by the municipalities would almost assuredly be passed on to those same homeowners through increased taxes or fees.

And then there were the wildfires. In 2017, California's fires wiped out premiums collected in previous years going back to 2001, according to a study by the RAND Corporation. Once again, there commenced a game of pass the hot potato. Insurers responded to the losses by dropping more than 235,000 homeowner policies in 2019, up 61 percent over 2018. After another disastrous fire season, in December 2019 the state imposed a moratorium on cancellations of policies for zip codes that covered more than 800,000 homeowners. In areas not covered, homeowners saw their insurance costs increase between three and four times, according to Amy Bach, founder of the nonprofit United Policyholders. After California's worst fire season in history in 2020, the state imposed a new one-year moratorium on cancellations in November 2020 that covered 2.1 million insured residences, or about 20 percent of the entire market. The year grace period was supposed to allow homeowners to fortify their homes against fires and/or find new coverage, but after the year, insurers were free to drop coverage. Doing that forces homeowners to buy the bare-bones coverage offered by the state's California FAIR Plan (CFP), a pool backed by all the state's insurers.

With desperate homeowners flocking to this last-ditch backstop each year, a crisis is brewing. The CFP cannot raise rates more than 100 percent at any given time. As more insurers pull out of the California market or drastically raise their rates, the prospect of a CFP bankruptcy becomes more likely should the state's catastrophic fire seasons continue. Watching all this carefully will be the banks, which are loath to take on uninsured or underinsured risk when writing mortgages. Watch home prices for an early indicator of a coming housing/banking crisis related to the fires.

On the opposite side of the country, in Florida, housing sales have already begun to fall for homes at risk of flooding. A study by the National Bureau of Economic Research focused on slowing sales as an early warning of housing price drops. In past housing crises, at first buyers are reluctant to pay up and sellers reluctant to drop prices, which results in slower turnover of housing stock. At some point the sellers blink, and prices begin to drop. The slowed sales began in 2013. As quoted in *The New York Times*, the NBER study attributes this inflection point to the shock of 2012's Hurricane Sandy, which awakened hundreds of thousands of northeasterners—major buyers of Florida real estate—to the perils of flooding. While prices and sales in areas of Florida at low risk of flooding continued to rise, both prices and sales dropped in high-risk areas.

Now insurers have to grapple with yet another threat of rising seas. As of this writing, it is still not clear what led to the horrific collapse of the Champlain Towers South condo complex in Surfside, Florida, on June 24, 2021, but insurers, not to mention government officials, inspectors, regulators, and, most of all, residents, now have to worry about whether ever more intrusive seawater is corroding foundations and changing the subsoil on which buildings have been constructed. Did changes in the composition of the ground underneath the tower contribute to its instability? Given the massive increase in construction along the coasts, the ubiquity of sea level rise, and the difficulties of examining the ground underneath completed buildings, this represents a monumental, if essential, task. The problem is that such an investigation needs to be done to determine the present safety of existing structures, and also their future viability as sea levels continue to rise.

There are other derivative impacts that amplify the economic damage of disruptive events such as climate change. To put it simply, both people and businesses will try to game or otherwise take advantage of the situation. Some insurers will use the increased frequency of storms to raise rates more than is justified by the risk. Some people will take advantage of an event like a storm to collect for damages that may not relate to the storm. Indeed, in what might be called a second-derivative impact, some insurers in Florida justified rate raises by claiming that rampant insurance fraud and frivolous lawsuits deriving from hurricane-related claims were threatening their very viability.

These shenanigans, however, camouflage the serious dilemma that derives from both California's fires and Florida's floods: if either risk was priced to the risks of climate change, insurance (and housing) would become unaffordable for large swaths of the public. There are legitimate social issues involved in keeping fire and flood rates low. Barry Gilway, the CEO of Florida's government-backed property insurance company Citizens, told the *Financial Times* that given the susceptibility of the Florida Keys to flooding, true pricing and/or flood-proofing of existing homes would make housing so expensive that the employees who service the area's economic engine—the tourist industry—could not afford to live there.

Similar stories can be found in every at-risk community. This is one reason that FEMA has been slow to update its flood maps to reflect the rising threat of climate-change-related floods. Flood maps compiled by the First Street Foundation estimate that 14.6 million properties are at risk for a hundred-year flood in the United States, nearly twice the 8.7 million properties identified through federal maps. If a house is in a federal flood zone and the owner has a government-backed mortgage (as most people do), they are required to have flood insurance. Many neighborhoods on the fringes of flood zones are occupied by people barely getting by, and the cost of flood insurance might make their homes unaffordable.

Analogous to the tug-of-war over flood maps is a struggle going on right now in California as it seeks to update its severe-fire-zone maps to more accurately reflect risk in an era of climate change. As is the case with flood maps, there are real economic consequences to changing these maps as the state imposes stricter building codes and other impediments to living in areas judged to be at severe risk. The combination of higher building costs and higher insurance rates would very likely make housing costs rise to unaffordable levels for many people who suddenly find themselves in a zone newly designated as severely at risk for wildfires.

Reality will likely settle this conflict between accurately pricing risk and accommodating the needs of those who live in at-risk areas. Between rising sea levels; increasing numbers of epic rainfalls; increased incidence and severity of windstorms; increased frequency, severity, and extent of wildfires; and increased temperatures and wet bulb temperatures, large swaths of the United States will likely become either unlivable or economically unviable.

An ambitious project published in September 2020 by the non-profit ProPublica and *The New York Times Magazine* used data and analysis from the Rhodium Group and other sources to project how climate change might transform the United States if it proceeded along the lines of the various scenarios compiled by the IPCC. The authors, Al Shaw, Abrahm Lustgarten, and Jeremy Goldsmith, compiled data for every county in the United States and assessed each county for its risk on a scale of one to ten for each of five different climate change impacts—temperature, wet bulb temperature, farm crop yields, sea level rise, and very large fires—and then came up with a number for economic damage for the period 2040 to 2060.

The resulting maps suggest that some of the most thriving counties in the United States will likely suffer extreme economic damage in the coming decades. Palm Beach and Miami-Dade counties in Florida had an economic risk of eight on the scale of ten. Beaufort County, South Carolina, home to the famed Hilton Head resort island, scored a nine on that same scale, thanks to a combination of future sea level rise and extreme wet bulb temperatures. A good number of the counties most at risk for economic damage had moderately high risks in a number of categories but scored high for extreme economic damage because of the anticipated compounding factors as risk stacked upon risk. There were also a number of counties, mostly in the Northeast, that will stand to benefit economically from these changes as climate refugees migrate northward.

The authors were wise enough not to put specific numbers on these predictions, but their basic points are well taken. If wet bulb temperatures create outdoor conditions that are unlivable, as is the prediction for ten parishes in Louisiana, or if outdoor temperatures are predicted to be so high as to be unbearable, as predicted for a number of counties in Texas, Arizona, California, and other states, or if much of your county will likely be underwater, or if farming is no longer viable, it is also likely that large numbers of people will pull up stakes and move. And when they move, their homes will be unsaleable, the mortgages will default, some banks will go under, and the tax base will disappear.

This scenario might have been avoided had the nation begun to adjust for the new risks when the threat was first recognized. And it still might be moderated were the United States to accelerate its adjustment to climate change risks going forward. To a large degree, however, many climate change risks are not priced in, sometimes for quite practical reasons (such as the need for affordable housing for local workers), making it ever more likely that when these risks surface, the eventual adjustments will be violent and disruptive.

Our modern market economy is so ingenious at spreading and hiding risk that its very adaptability has become a threat to its future. Our society is so good at monetizing discontents (think MAGA hats) and finding profit opportunities that its very adaptability has become maladaptive. We are so gifted at finding the profit to harvest in every risk and at pushing off the day of reckoning that, as a society, we have lost the ability to recognize and adjust to true danger. The role of politics and the insurance industry in relation to global warming offers a case study of this paradox in real time. The net effect of slicing, spreading, and camouflaging the risks of fire, flood, drought, and heat attendant to climate change has been to maintain the status quo and reduce any impetus to adjust to these threats.

The 2010s saw the business community begin a momentous shift toward weighting the costs of climate change over the costs of efforts to moderate its effects. It also saw the moneyed interests dramatically increase their investments in clean tech and other opportunities climate change might offer. In this sense, the clock of business and finance, which used to lag the public, the scientific community, and reality, is now running ahead of the public and just behind the scientific community. Since money drives the system, it's a welcome trend. But can it catch up fast enough to make a difference in our future?

PART SIX

Where Do We Go from Here?

Nineteen

THE TRAP WE'VE SET
FOR OURSELVES

The quality of life for hundreds of millions if not billions of people, the very possibility of life over the next few decades, depends on how we confront changing climate and our role in those changes. The odds are stacked against us. At the moment, the possibility of a climate catastrophe might be likened to two express trains racing toward each other on the same track. One train is propelled by our politics and economy, which gives business as usual powerful momentum. The other train is the momentum of climate change itself, which intensifies and accelerates as we pump more and more greenhouse gases into the atmosphere. As the metaphor suggests, the time during which we might avert this train wreck shortens rapidly as these two trains accelerate and converge. Still, the train wreck is a probability, not a certainty.

Here are some of the positives:

The most promising trend has been the discovery that vast amounts of money might be made and millions of jobs created in the fight to avert this coming catastrophe. A section on the new renewable supermajors published in Bloomberg's *New Energy Outlook* estimated that $11 trillion in investment will be required in renewables over the next thirty years. Bloomberg estimates, however, that even with that investment temperatures would still rise by 3.3 degrees Celsius by the end of the century. This would be ruinous, but perhaps it is still avoidable. As Steve Pacala noted, development and adoption of clean energy has proceeded far more rapidly than he anticipated little more than a decade ago.

Other developed (and some developing) countries have moved away from fossil fuels far more rapidly than the United States. More than half of Germany's grid capacity now comes from solar and onshore wind power. Around the world there have been periods where provinces have met electricity needs without any reliance on fossil fuels. As fossil fuel projects become increasingly difficult to finance (and renewable projects easier to finance), and as rechargers expand their reach and increase their speed, and as battery technology improves, and as renewable energy storage becomes easier, and as tidal, deep geothermal, and other renewable projects proliferate, this transition might accelerate and perhaps stop the rise in emissions earlier than even the optimistic forecasts.

Stopping the rise in emissions won't be enough, of course. Global temperatures respond to the levels of greenhouse gases and not the rate of change. Reducing levels of greenhouse gases without unleashing severe unintended consequences is a tricky business, but there are several benign, even environmentally positive strategies to suck carbon from the air at the scale needed to make a difference. Perhaps the most benign would be restoring wetlands, reducing farm emissions, and, most important, halting deforestation and reforesting degraded lands. A massive study led by Bronson Griscom, a forest expert at Conservation International, and published by the *Proceedings of the National Academy of Sciences* in 2017 estimated that such "natural climate solutions" could cost-effectively sequester greenhouse gases equivalent to between 11 and 15 billion tons a year, or between a quarter and a third of annual human-sourced emissions.

Then there are the industrial solutions. One of the more intriguing efforts comes from a start-up named Blue Planet, which, through a proprietary process, combines calcium and CO_2 to make synthetic limestone that in turn can be used to make concrete. If adopted, this substitution has the potential to pull tens of gigatons of carbon out of the air each year. Norway has greenlighted a similar project to sequester carbon at industrial scale.

There are other scalable, financeable proposals for carbon capture. Between reducing fossil fuels usage, increasing renewables, and recapturing CO_2 already in the atmosphere, it is possible that we might bring down the amount of GHGs in the atmosphere. As the global temperature record suggests, *all things being equal* (important caveat), climate seems to respond quickly to levels of greenhouse gases.

Preindustrial levels of CO_2 were 280 parts per million. Global temperatures remained near the long-term average (based on a five-year smoothed average) until the late 1970s, when carbon concentrations reached 340 parts per million, or about 20 percent higher than the preindustrial average. Temperatures probably would have begun rising sooner, but after World War II, a massive increase in sulfate aerosols in the atmosphere, a result of the postwar boom and major volcanic activity, had an offsetting cooling effect that lasted until the 1970s. Since 1980, temperatures (again, on a five-year smoothed average basis) have been rising pretty much in lockstep with greenhouse gas emissions. If temperatures are that responsive to CO_2 on the way up, there is hope that they will be equally responsive on the way down—unless, on the way up, we breach the thresholds that trigger out-of-control, self-reinforcing warming.

Collectively, we must all hope such symmetry applies and that thresholds are not crossed, as the consequences of a self-reinforcing warming would exceed the death and misery resulting from a global nuclear war. This is not hyperbole.

Humanity has enjoyed a ten-thousand-year honeymoon under the climate regime that has prevailed since the end of the last ice age. During that period our numbers have expanded from a few million to 7.8 billion. Civilizations have arisen, and we have expanded our cultivation practices over most arable land and into every ecosystem, including the oceans. To use an analogy from finance, humanity has leveraged its position far beyond what is ecologically sustainable based on the assumption that the benign climate that nurtured our dominance would continue.

But climate is changing. As Stefan Rahmstorf noted, we have left the Holocene, that benign hiatus during which we have been fruitful and multiplied. Carbon dioxide levels are now what they were during the Pliocene, roughly four million years ago. This was before the ice ages began. Temperatures back then were 2 to 3 degrees Celsius warmer than they are now. The Greenland Ice Sheet had yet to form and sea levels were about 80 feet higher. There was plenty of life on the planet, and plenty of apes too, but there were no humans. That's where we're headed on our present path.

We know where we're going, we know what the likely consequences are, and we have the tools to temper, if not avert, a coming catastrophe. The existential question: Will we?

The "we," of course, is humanity, which includes a few hundred million people deeply concerned about global warming, several hundred million people well aware of the dangers ahead, as well as several billion people who are either unaware or only vaguely aware of the threat. Fortunately, those several hundred million people live in developed societies that account for most global emissions and have the technological sophistication to reduce emissions and recapture greenhouse gases already in the atmosphere. Unfortunately, as yet, this group is neither sufficiently alarmed nor sufficiently organized to do what will be necessary. Recall that the Climate Protection Index that ranks nations for their efforts to combat the threat has consistently left the top three spots open because, in the opinion of the groups that contribute to the index, no country has done enough to avoid dramatic warming. And as recent history has shown, the decision by China and India to embrace coal to power their development in the 1990s proved more consequential for climate than the patchwork of many thousands of greenhouse gas reduction initiatives launched by cities, states, provinces, and nations in the developed world during that same period. And thanks to a release of documents by Greenpeace, we discover that as late as October 2021, influential nations such as Japan, Australia, and India were still trying to water down language on the transition away from coal and fossil fuels in submissions to an important IPCC meeting.

What may prove to be the fatal flaw of our modern market economy is that it can't do collective action. Our tightly coupled global economy fosters the immediate transmission of bad things such as computer viruses, actual viruses, and financial panics. But our responses to any of these outbreaks is typically fragmented and slow.

The most recent crisis—the COVID pandemic—crisply illustrates the limitations of the current economic/political paradigm. Even with immediate life-and-death consequences, our current global order is not configured to respond to a global crisis. The virus somehow jumped to humans in Wuhan, China, and then exploded because Wuhan officials at first refused to acknowledge the outbreak. The United States was slow to react because the Trump administration had ended funding for a pandemic early-warning initiative called PREDICT and two years earlier dismantled a directorate on pandemics in the National Security Council. The Trump administration consistently downplayed and politicized the pandemic, allowing cases and deaths to explode in the United States. Commonsense measures such as mask wearing and social distancing were viewed as assaults on liberty and freedom by significant percentages of the population.

The result of this politicization has been near utter failure in the United States to contain the pandemic relative to other countries with reliable health statistics. The U.S. population, for instance, is a little more than 2.5 times the size of Japan's but has had roughly ten times as many deaths from the disease and nearly one hundred times as many cases (as of this writing). The United States is six times the population of South Korea but has had (as of this writing) about four hundred times the number of deaths. The United States is thirteen times the size of Australia but has had close to four hundred times the number of deaths. Adjust for demographics and comorbidity factors all you want—there remains an astonishing gap between the performance of the United States, the world's dominant economy, and dozens of nations that have been somewhat successful in keeping the disease under control. Of course, the United States was not alone in failing to contain the pandemic. Russia, Brazil, Mexico, India, South Africa, France, Spain, the United Kingdom, and a number of other nations have also seen the virus rip through their populations.

Piecemeal does not work with a global problem. Even if a nation has eliminated the virus, it will still suffer economically so long as the virus is rampant elsewhere, disrupting travel, economic activity, and supply chains. The implications of the COVID crisis for the climate crisis are obvious. Only collective action can resolve either one. If the global community cannot come together to deal with a disease that is killing and debilitating people on a scale not seen in the developed world in a hundred years, how can it come together to deal with an atmospheric threat that for most of the world's population is either unknown or who believe that the danger lies off in the future?

In the United States, the response to COVID suggests that the nation is not ready to grapple with any global problem. Perhaps the only thing that might be less subject to politics than a virus might be physics, but as we have seen, both the virus and climate change have become completely politicized. The rise of partisanship has reversed more than fifty years of progress in terms of American attitudes toward the environment.

This realization has shaken me to the core. It used to be said that environmental awareness has become an American value. Polls have shown consistent support for environmental initiatives.

The rise of intense partisanship may partially explain voter behavior and attitudes toward climate change, but a broader question involves the decline of critical thinking on issues of importance. After the 2020 election, Trump had large swaths of the Republican electorate believing a bizarre conspiracy about Dominion voting machines. The accusation was that Dominion was switching votes, even in states governed by pro-Trump Republicans and even where the voting machines were backed up by paper ballots. He continued to push this conspiracy and his supporters continued to believe it, even after the Republican secretary of state in Georgia finished a recount comparing the paper ballots with the voting machine results and found no meaningful difference.

And then, of course, his backers invaded the Capitol on January 6, 2020, in an attempt to disrupt the counting of the Electoral College votes even as it was proceeding. So, even though Trump's arguments about a stolen election fell apart upon the slightest scrutiny, defied common sense, and had been rejected in more than sixty court cases, a few of which reached the Supreme Court (which is dominated by conservatives and judges he appointed), thousands of his followers were moved to invade the seat of U.S. democracy, a place whose sanctity had been violated by a mob only once before, more than two hundred years ago. Moreover, this insurrection was supported by about 45 percent of Republicans according to a poll conducted by YouGov after the event.

Susceptibility to conspiracy theories has long been a subject of fascination and research for political and social scientists. As the mainstreaming of truly absurd conspiracy theories such as those espoused by QAnon demonstrates, this susceptibility has only grown in recent years and has now reached a point where it interferes with public policy, legislation, and the democratic process itself. Technological change has enhanced this susceptibility, which means that our elected leaders are going to have to solve the problem of a public passionately clinging to "alternate facts" even as they try to act on climate change. This does not bode well for our future.

One factor that has left people vulnerable to conspiracy theories has been the decline of traditional authorities. Trust in academia, policymakers, and the scientific community has steadily eroded, an alienation exacerbated by income inequality and the growing cultural divide between the educated elite and an increasingly tattered middle class. Trump accelerated this process with his attacks on expertise and his appointments of quacks and hacks to positions that formerly required legitimate credentials. With critiques from traditional authorities drained of their authority, preposterous ideas persist and fester. There's a deep need in people to attach themselves passionately to ideas and structures larger than themselves. A Trump rally had more the flavor of a religious revival than a political event, and when people went all in on Trump, they also went all in on his articles of faith—one of which was that COVID was fake news. Many would cling to those articles of faith even as it was costing them their lives because the belief that COVID was a hoax had become a key part of the integrated Trumpian worldview, and to abandon that belief would undermine all the other fictions they had bound themselves to.

The rise of the internet and social media has been matched by the decline of intermediaries in all fields, including finance, politics, and, perhaps most of all, the media. The intermediaries in the media were the editors, producers, and fact-checkers who previously provided a tether to a common set of facts and also friction so that unsubstantiated fabrications wouldn't gain traction before they were vetted.

Now, on the internet, fact-checkers have disappeared, as have most editors, all news sites *look* equally authoritative on a screen, and people can choose news sources that bolster the isolation of whatever silo they choose to inhabit. Newsmax, a peripheral, far-right news source that became the go-to place for those who think the 2020 election was stolen, went from just over 58,000 prime-time viewers on its obscure cable channel to more than 1.1 million viewers for its most watched shows in the weeks after the election, according to Ben Smith, the media critic for *The New York Times*. Those ratings are better than all but the very most popular mainstream news shows on cable. Christopher Ruddy, the founder of Newsmax, told Smith, "In this day and age, people want something that tends to affirm their views and opinions." And for all the thundering about liberal bias in the media, Fox had four of the top five cable news-related shows in the weeks after the election.

How likely is it that climate change will shed thirty years of political spin and disinformation? It's true that at least some of Rupert Murdoch's children recognize the seriousness of climate change. Perhaps they will exert pressure on the Fox network to abandon its divisiveness on the issue. As the example of the 2020 election showed, however, when Fox News recognized the reality of the Biden win, a significant portion of their audience vilified the network and sought out outlets such as Newsmax that still reinforced the delusion that the election was stolen. If Fox changes its tune on climate change, there will still be plenty of alternative news sources that reinforce the notion that global warming is some sort of hoax.

Daunting as is the prospect of a collective recognition of the seriousness of the threat, there is an even more formidable impediment to collective action: the very nature of our modern market economy. The late Charles Reich, author of *The Greening of America*, once remarked to me that the corporate elites don't run the "system" (the word for the establishment very much in vogue in the late 1960s when Reich made that remark), they "tend" it. That observation was subtle, and brilliant. Over the decades, our consumer society has become optimized toward identifying opportunities for profitable investment as well as hedging, spreading, and shifting the risks entailed. Those who contribute to profits and defanging risks get rewarded; those who don't eventually find themselves marginalized or shed.

The system is amoral. If a cohort of ethical investors shies from a particular area—porn, cigarettes, coal—earnings ratios fall, offering fat returns for investors who care for profits more than probity. So long as there is consumer interest or a market anywhere and a product can be profitably produced, front-line agents will continue to sell it until they can't. And so long as there is profit to be made, an industry's lobbyists, lawyers, and other support systems will try to protect it.

This gives established businesses tremendous inertia. Once a profit engine is identified and infrastructure has emerged to support it, it becomes entrenched in the economy and it takes enormous energy to dislodge it. Recall that the global community took action on CFCs only when there was "smoking gun" proof that they posed a threat to life on earth, several years after the ozone hole was first discovered. Even then, action might not have been taken had not the dominant producer, DuPont, also had the lead in producing alternative refrigerants, which made it in their interest to ban CFCs.

If the CFC business is a molehill in the global economy, the fossil fuels complex is Everest. The momentum of this colossus is infinitely greater than that of the CFC industry. Oil, coal, and natural gas are entrenched throughout the modern economy, with trillions invested in infrastructure, supporting millions of jobs, and protected by a massive network of lobbyists and politicians. Fossil fuel's power and persistence is evident in the fact that the overwhelming majority of the world's vehicles still rely on an engine that was invented before the Civil War—the internal combustion engine—to power all manner of vehicles even as we enter the third decade of the new millennium.* Yes, an accelerating transition to clean power and EVs is underway, but we shouldn't discount the truly massive numbers of workers, investors, bankers, and managers, not to mention local, state, and national politicians, who have a vested interest in continuing the use of fossil fuels.

If the system is amoral, it is also blind. Its genius at organizing itself to exploit profit opportunity wherever it arises is matched only by the ingenuity with which it hedges, spreads, and shifts risk. The ace in the hole is that last point—the *shifting* of risk. As discussed in earlier chapters, the ingenuity of the insurance industry and the wizardry of Wall Street financial engineers have been very successful at delaying any reckoning for the risks of climate change, which have been surfacing as steadily mounting claims for flood, wind, fire, and drought damages from natural disasters. In the ultra-low interest rate environment that has prevailed for the past dozen years, there will always be investors, looking for higher yields, who will buy a CAT bond paying 8 percent while crossing their fingers that a hurricane won't hit Miami in the next three years. And if that market dries up due to a surge of catastrophes, there's the ultimate ace in the hole— the taxpayers of the United States. Once risk has been spread so wide that it ends up with the nation's taxpayers, it is also effectively camouflaged, and camouflaging risk allows business as usual to continue long after financial prudence would dictate changing course.

The net effect of spreading risk is to allow it to accumulate. To

---

* Even in 1910, Thomas Edison recognized the limitations of fossil fuels, saying in an interview with an influential publisher, "This scheme of combustion in order to get power makes me sick to think of—it is so wasteful.... Sunshine is a form of energy, and the winds and the tides are manifestations of energy."

repeat, if a risk is real, it can't be destroyed through financial wizardry. And if a risk is real, each year it isn't expressed increases the probability that it will be expressed in much larger terms in the future. For example, wildfires are part of the natural ecology of the West. Suppressing smaller fires, as most states did for decades, allowed massive buildups of fuels and tinder. Then, as climate change raised temperatures, dried out the air and soil, and increased the intensity of Santa Ana winds, much of the West became a tinderbox, and the probabilities of a major conflagration rose to a certainty. So too with the financial risks of climate change. To the degree to which spreading risk to the taxpayers underprices insurance for homes and businesses in fire, flood, and other zones vulnerable to the changes wrought by climate change, the current system allows people to be blind to the risk of global warming and guarantees that these risks will become systemic.

If we actually lived in a true free-market libertarian paradise, it would be horribly ill-equipped to deal with a collective-action problem like climate change, but at least risks would be borne by those who were paid a lot to take those risks. Some would go bust, survivors would raise their prices, and people would adjust. Effectively, this would be the equivalent of a prescribed burn of western brush. That doesn't happen anymore in the United States.

Consider 2008, when the housing collapse rendered much of our financial system insolvent. Faced with systemic risk, the federal government stepped in with bailouts, effectively protecting shareholders, bondholders, and management from bearing the risks they had been handsomely paid to assume. In a classic act of the barn door closing after the cows have fled, Congress passed the Dodd-Frank Act, restricting some of the more irresponsible risk taking of financial institutions and setting up the Consumer Financial Protection Bureau (CFPB) to give some voice to the millions who didn't get a bailout—a group that included the homeowners and consumers forced into bankruptcy by the crisis.

Then lobbyists began picking away at Dodd-Frank and Trump gutted enforcement actions by the CFPB, setting the stage for another multitrillion-dollar bailout as COVID caused a panicked deleveraging in the funding markets in March 2020. The overnight funding markets that provide liquidity for much of the banking system dried up, once again threatening systemic risk. The need for financial system bailouts every ten years caused Minneapolis Fed president Neel Kashkari to describe our financial system as "absurd" in a September 2020 speech to the Council of Institutional Investors.

The socialization of risks related to climate change has proceeded differently. The camouflaging of these risks is the result of an accumulation of different backstops implemented over the years in response to a wide variety of threats. Federal subsidies come through the deficit of the National Flood Insurance Program; through the Disaster Relief Fund, which provides assistance for declared emergencies; and through separate programs for various agencies such as the $3 billion appropriated for the USDA to cover crop losses related to natural disasters. Then there are the state and local programs. Apart from the backstop state-sponsored insurance programs in Florida and California, most states have disaster relief programs or other ways of socializing risk.

Regardless of how those programs are structured, the ultimate backstop for those risks is the American taxpayer. California, for instance, is either going to have to relieve the bankrupt giant utility PG&E of some portion of future liability to wildfires, buy the utility, or allow the utility to increase rates. In every single case, climate change liability ultimately lands on customers, taxpayers, or both.

In 2019 in the United States, 52 percent of weather-related losses were insured. State and federal programs closed some of that gap, and they will increasingly bear that burden in the future. Figures from the Peterson Institute show that weather- and climate-related losses since 2010 were 56 percent higher than the previous decade and nearly three times losses in the 1990s. These numbers will continue to grow as climate change worsens, putting an even greater burden on budgets at all levels of government. Socializing climate risks has enabled the United States to kick the can down the road, but each year thus kicked, the can gets bigger. At some point, perhaps soon, it will become too big to kick anymore.

Our current system is amoral, blind, and, not least, it can be gamed. A perfect example came from the response to the COVID crisis. Congress gave $670 billion to the Paycheck Protection Program in the first round of stimulus to provide forgivable loans that would protect jobs for small businesses. Subsequent analysis showed that most of that money went to large businesses, that many of the loans protected only one job, and that significant funds went to billionaire-owned businesses and even private-equity firms. Unable to get funds, tens of thousands of small businesses went under, while the stock market soared. The fundamental problem was that grants didn't go to the most in need, they went to those companies whose managers and lawyers were adept at filing forms and/or had the connections to speed the applications along.

The problem of a system ripe for gaming is that funds directed at a problem often don't get to where they're needed. As the example of CDM funds going to build a coal-fired electric plant in India illustrates, this is not just a U.S. problem. One wonders, for instance, to what degree the push for building coal-fired plants throughout Asia was propelled by off-the-books "incentives" at all levels of these heavily state-controlled economies. The susceptibility of federal programs to succumb to either bureaucratic inertia or perversion of their mission through gaming is perhaps the most credible of all the libertarian critiques of big government.

What we have right now is a blind, amoral system that invites gaming and manipulation by the clever. It is a system that preserves entrenched interests and camouflages risk, reducing the motivation to adapt when problems appear. It is a system whose default is to drive off cliffs. It is not the system you want when confronted by an existential threat that demands extremely rapid adjustments to the sources of energy that power the economy. But it's the system we have.

"We have a system that is so supremely adaptable that it has lost the ability to adapt to true danger." That's a sentence that I wrote in 1980. Now true danger is upon us. The most likely path is that we will drive off the climate cliff.

Through our ingenuity, adaptability, and greed, we have created a trap for ourselves. All too many of us remain oblivious to the danger. The next and final chapter will offer some thoughts on how we might yet escape.

Twenty

A NARROW PATH TO
A LIVABLE FUTURE

I spent some years as chief investment strategist for a hedge fund that specialized in investing in distressed companies—businesses that were circling the drain and were either bankrupt or dangerously close to going under. In that role, I served as director of a few of such companies once they were reorganized. I remember one where the board would regularly set goals for the CEO that he would have to meet to get his bonus. And I remember that he regularly achieved those goals—right up to the moment we once again filed the company for bankruptcy. This is how I view international efforts thus far to prevent catastrophic global warming. Nations might meet their goals, and we will still face a climate catastrophe.

The world has begun decarbonizing, but we need to accelerate the process greatly to avoid a climate disaster. The Green Deal announced by the European Union would have the political unit that represents the third-largest source of greenhouse gas emissions become carbon neutral by 2050. An aggressive goal, but consider what that means: nearly thirty additional years pumping more greenhouse gases into an already overburdened atmosphere. Even if the world met the terms of the Paris Agreement, we face further warming that would bring humanity into uncharted territory. A 3 degrees Celsius or more rise in global temperatures would produce a world hostile to farming, humans, and countless other creatures. This is where we are headed. This cannot happen. Stronger measures than far-off dates for carbon neutrality are needed to avert calamity.

It's not just the United States that must drastically reduce its emissions, but most other developed nations as well. Moreover, emerging nations need to find a path to development that does not overwhelm efforts to decarbonize by other nations, as has happened over the past two decades during the industrial development of China, India, and other new economic powers. The atmosphere doesn't care where emissions come from; it reacts to the level of greenhouse gases it bears. We need a tool that will create very strong incentives in every country to reduce net carbon emissions. We have that tool, though it is anathema to subscribers to the neoliberal economic order of free trade and open competition that has dominated globalization since the fall of the Soviet Union.

It's called tariffs.

The previous chapters have tried to show the tremendous momentum of business as usual, regardless of the economic system. In the United States, the spreading and socialization of the risks of climate change tend to deflate any sense of urgency about the issue. Russia, a kleptocracy, has leaders who openly scorn the threat of global warming. So does the president of Brazil, Jair Bolsonaro: his tolerance of the illegal burning of the Amazon makes Brazil the world's largest contributor of GHG emissions coming from deforestation. China, a state-managed economy, is now the largest purchaser of renewables, but it is also the largest emitter of greenhouse gases. And everywhere, even in nations with solid national policies, people cheat.

Here's how climate tariffs would work. It is now possible to monitor greenhouse gas emissions by point of origin using satellite-based remote sensing technologies, and even more refined monitoring is on the way to meet the requirements of the Paris climate accords. That data could set a baseline of the greenhouse gas emissions of 194 nations. Then a percentage goal for annual reductions in emissions could be established following a short phase-in period, with tariffs adjusted based on success or failure to meet those goals (outstanding success should be compensated with credits). Following data collection, a nation might be given a warning and one year to cure the deficiency before tariffs are imposed.

The key is that the tariffs would be universal, with one set level for every nation on earth. Don't get cute. As the history of climate negotiations has shown, specificity and exemptions invite gaming, endless negotiations, and cheating. Mali can achieve a 3 percent reduction from its base, for instance, just as easily as the United States from its base. Renewable energy now competes with or beats any fossil fuel on costs, and there are now many ingenious ways of reducing capital costs. New, scalable carbon capture technologies are coming online.

Nations could choose their own paths to compliance. European countries might focus on EVs and renewables, a continuation of what they are doing. Some nations might want to put a price on carbon. Sweden, for instance, prices carbon at about $130 a metric ton—four times the highest price yet reached in the EU carbon markets. Brazil might dramatically lower its emissions simply by controlling illegal deforestation in the Amazon, something it should be doing anyway. Indonesia could take a similar path, and its growing manufacturing sector would have a strong interest in making sure the country did not pay tariffs.

An across-the-board tariff creates an incentive within a country for competing interests to police bad actors. Tariffs at the national level would solve the free-rider problem that has dogged global warming initiatives from the beginning. Because of advances in renewables, EVs, and efficiency technologies, compliance on emission reductions will be easier than at any previous time in the climate change era, and it will be easier still as the buildout of renewable infrastructure accelerates.

Such a tariff regime could be created under a revised Paris Agreement in coordination with an existing international forum such as the World Trade Organization (which in its report *The WTO at Twenty* recognizes that climate change is an issue it must address). Tariffs could be collected by the importing nation and then be pledged to an international finance institution such as the World Bank, and the money could then be allocated to clean development projects directly targeted to reducing emissions in the poorer states. The fact that every nation has to meet its emission targets or be subject to tariffs would help prevent the counterproductive diversion of money that characterized prior international attempts to fund clean development.

The Trump administration's retreat from international obligations actually made it easier for a climate tariff regime to be implemented. Even though Biden has made a priority of reengaging, the Trump years gave other nations the opportunity to think about an international order without U.S. leadership. Emboldened nations, such as France, have already threatened to impose tariffs on some U.S. goods because of our retrograde climate actions of the past four years. A global problem such as climate change demands collective action, not a piecemeal approach. So far, the collective actions tried have been all carrots but no sticks. And, demonstrably, they haven't worked.

Tariffs could provide that missing stick. I expect the proposal will horrify economists because of the long history of beggar-thy-neighbor tariff case studies. It should be clear, however, that this would not be the case with a universal tariff. The regime would not be one nation seeking advantage. It would provide incentives for every nation, and its goals would be achievable. The ideal implementation would have its administrators twiddling their thumbs because all nations found compliance both affordable and in their interest.

A universal tariff might be our only hope to reduce GHG emissions before disaster strikes, but the deeper problem is the skewed incentives of our consumer society that render our economy amoral, blind, and easily gamed. If we imagine the consumer society as an amoeba, changing its shape to feed off anything that smacks of profit potential, those incentives are its sensors. In other words, perverse incentives are intrinsic to the functioning of our consumer society. At a time when we need an economic system that recognizes and adapts to hazards, we have one whose sole impetus is to maximize profit.

As we've seen with the extraordinary speed in the development of a COVID vaccine, the incentives of a consumer society can be channeled toward solving problems. The tens of billions directed at the vaccine effort by government worked in part because developing a vaccine was not a threat to business as usual for any part of the supply chain, just an extension of what they were already doing. Averting a climate catastrophe remains a threat to an immense part of the economy, but that is changing rapidly as the move away from fossil fuels accelerates. The day might be approaching where the potential for profit from the shift might overwhelm the diminishing vested interests in the status quo.

That might provide some comfort if climate change was the only danger facing humanity. Instead, almost every ecosystem on the planet is nearing the brink of collapse as our consumer economy attempts to satisfy the rising expectations of billions of people. "Solving" the climate crisis—a very long shot in itself—does not let the current system off the hook. There are simply too many other crises hurtling toward us. The collapse of fisheries, competition for fresh water, the prospect of more pandemics as pathogens jump from plants and animals to humans, wholesale shifts in rainfall patterns as tropical rainforests are cut, the disappearance of pollinators, and so many other threats will continue to worsen regardless of whether climate stabilizes.

The consumer society is the problem. Our economic system is unsustainable, given the combined impact of human numbers, human aspirations, and the way we treat the biosphere. While this book has focused on the decisions and missed opportunities that brought us to our present predicament with regard to climate change, a look forward involves the broader question of whether we can reorient our economy toward a more sustainable footing. I'll offer two likely paths for the future, one of which points to a way forward and the other a new Dark Age.

The consumer society can't change or reform, but one of its instabilities provides a view of what might replace it. Part and parcel of the profit optimization of the modern economy has been the widening gap between rich and poor. This is a trend that has been accelerating since the 1980s, and it has been exacerbated by the loss of labor bargaining power resulting from corporate access to a global labor supply, and as automation has proceeded from routine industrial tasks to expert and managerial systems. We're fast approaching a society where those with capital, the technologically adept, networkers, some essential workers, and a few leaders prosper, while the great mass of humanity finds itself with few prospects and no bargaining power.

If one aspect of the consumer society is that it is ecologically unsustainable, the winner-take-all nature of a consumer economy makes it politically unsustainable. For all its momentum, the consumer society is fragile. In a democracy, the few can hold on to their privileges only with the consent of the many. A significant portion of Trump's 2016 support came from those who felt that the system was rigged against the working person (given what Trump did once in office, this ranks as one of the greatest examples of bait and switch in American political history). Evidence of this distrust was that in 2016 about 15 percent of Bernie Sanders supporters switched to Trump rather than to Hillary Clinton. As the profit-optimization imperative of a consumer economy slowly strangles the prospects of average Americans, this trend toward distrust of the elites will intensify.

The dark path is one already evident in Trump's legions, who deny the reality of the 2020 election and proved willing to violently overthrow the will of the people and more than two hundred years of precedent, and to invade the very seat of democracy to preserve his power. It's evident in the antidemocratic leaders who've come to power in democracies elsewhere such as Orban in Hungary, Erdogan in Turkey, and Bolsonaro in Brazil. Should these forces prevail in the United States, it's difficult to imagine how the union would hold together, much less meet the multiple challenges looming over our future.

History has shown us how societies react to unstable times. Typically, people turn inward. Xenophobia rises. People take out insurance in the form of strengthened family ties and strengthened affiliations to neighborhood, cultural, and religious groups. There is less innovation and investment, as well as less social experimentation. It's not all bad, but a society wracked by internal conflict is not one that might address external threats.

Then there's the other path. Libertarians and small-government conservatives are not going to like it, but it is the natural response to the trends of income inequality we have seen for several decades. If the voters finally have had enough with a winner-take-all economy, they might choose candidates advocating for the democratic socialist model that characterizes most of the Scandinavian countries of Europe as well as Germany, France, Italy, and the United Kingdom (though my suspicion is that if the United States moves in that direction, somewhere along the way, the word "socialist" will disappear). All these nations have far stronger worker protections and far more robust safety nets. Germany, France, and numerous other European countries have varying degrees of worker representation on boards of directors. Most of the nations that rank higher than the United States in terms of quality of life as determined by the World Economic Forum have a democratic socialist political system.

A counterreaction to the unfettered capitalism that the United States has aspired to (but not practiced during crashes) would also likely result in a government that would be more willing to use the tax laws for social and environmental goals. Between the tax laws and regulations, government might be able to recognize and address long-term threats that we currently ignore. The democratic socialist approach is anathema to libertarians, who have been fighting tax laws for decades, but a world in which the depleted middle class uses its voting power for constructive reform would not be a world where libertarian ideas still prevailed.

Of course, a voter reaction against wealth inequality and the wage gap might not stop at a moderate solution such as democratic socialism. A movement to the extreme left could be as destabilizing as a movement to the extreme right. Keep in mind that the 2018 yellow vest movement in France, a populist uprising that drew from the financially stressed across the political spectrum, was originally prompted by a rise in fuel taxes that was in part intended to address climate change.

There is no guarantee that it will be something better that replaces the collapse of an unsustainable system, but there is a guarantee that an unsustainable system will collapse. It behooves us to think about what will replace it once it does. The threats facing society are of such scale that they will require well-functioning governments and the support of the people if they are to be addressed. The diamond magnate Harry Oppenheim once remarked, "If they don't eat, we can't sleep." This is something the privileged might well keep in mind as our consumer society hurtles to meet its appointment with its own contradictions. A movement toward democratic socialism offers a plausible off-ramp. There may be others, but time is not our friend.

Acknowledgments

Fire and Flood is the product of more than three decades of writing and thinking about climate change, but recent interviews and conversations with a few people proved crucial as I tried to reconstruct the events that have led to our present situation. Conversations with Anthony Leiserowitz, Amory Lovins, Stephen Pacala, Thomas Johansson, Gus Speth, and Naomi Oreskes helped me fill out the picture on such issues as the flux of public opinion, its intersection with politics and policy, the interplay between the developed nations and the big emerging economies, and the technological progress of alternative energy. I've been to many conferences on climate change and the insurance industry over the years, but numerous conversations with Chris Walker and Frank Nutter helped me understand the industry's evolving response to the threat. My brother, David Linden, who owned an insurance agency for many years, helped me understand the incentives at the retail end. A sixteen-year stint as chief investment strategist at Bennett Management, a family of hedge funds focused on distressed situations, enriched my understanding of the financial markets, particularly the various connections among the housing market, the banking system, and the larger economy, a vulnerability that came vividly to the surface during the financial crisis of 2008.

With regard to producing the book, this ranks among the most pleasant publishing experiences of my career (so far and knock on wood). At Penguin, Scott Moyers offered sound advice on how to bring to the foreground the major themes of the book. Then he followed up with a deft and elegant edit, which cleared away some of the eddies that impeded the flow of the book. Throughout the process, Scott displayed relaxed good humor and made me feel at home. Scott's associate, Mia Council, provided a steady hand, navigating the manuscript through the publishing maze. Esther Newberg provided her usual superb representation, and her assistant, Estie Berkowitz, somehow kept a straight face while rescuing me from some of my amateurish screwups while formatting the manuscript. For fact-checking I took a chance on a talented high school senior, Charlie Rudge, and he came through brilliantly (thank you, Vladimir Klimenko, for recommending Charlie). Fact-checking science when the science is rapidly changing is no easy task, but Charlie proved up to the job. David Bjerklie, who has fact-checked my work going back to my years at *Time*, stood ready to back up Charlie, but, largely unneeded in that capacity, used his peerless nose for identifying trouble spots. Amy Ryan did an outstanding job copyediting the manuscript. With all these eyes looking for pitfalls, any pits I fall into are of my own making.

This book is built upon a perspective on climate change developed over more than three decades writing about the subject. The perspective is mine, but over those years I have benefited from generous access granted by scores of researchers—too numerous to name here—coming from the spectrum of climate sciences, as well as from economics, epidemiology, archaeology, anthropology, and even paleontology. A number of institutions have generously facilitated my access to their researchers and research. Thanks to all!

Suggested Further Reading

No human could keep up with the ongoing flood of scientific studies, corporate and think tank reports, general interest articles, and books on climate change. The sources cited in the text of *Fire and Flood* offer a first cut of significant papers, studies, and analyses should readers want to further explore the complex ganglia of global warming. The following list of suggested readings is short and sweet and compiled for the nonexpert reader who is interested in delving deeper into the history of the science, the complicated history of humanity and climate, the battle to address the threat, and what it might mean for the future. A number of these books have been updated, so look for the latest editions. All are highly readable.

The Development of Climate Science

Alley, Richard. *The Two Mile Time Machine: Ice Cores, Abrupt Climate Change, and Our Future*. Princeton, NJ: Princeton University Press.

Broecker, Wallace. "The Carbon Cycle and Climate Change: Memoirs of My 60 Years in Science." *Geochemical Perspectives* 1, no. 2.

Kolbert, Elizabeth. *Field Notes from a Catastrophe: Man, Nature, and Climate Change*. New York: Bloomsbury.

Weart, Spencer. *The Discovery of Global Warming*. Cambridge, MA: Harvard University Press.

Climate Change, Weather, and Human History

Caviedes, Cesar. *El Niño in History: Storming Through the Ages*. Gainesville, FL: University of Florida Press.

Davis, Mike. *Late Victorian Holocausts: El Niño Famines and the Making of the Third World*. Brooklyn, NY: Verso.

Lamb, H. H. *Climate, History and the Modern World*. London: Routledge.

Linden, Eugene. *Winds of Change: Climate, Weather, and the Destruction of Civilizations*. New York: Simon & Schuster.

The Battle for the Public's Attention

Conway, Erick, and Naomi Oreskes. *Merchants of Doubt: How a Handful of Scientists Obscured the Truth on Issues from Tobacco Smoke to Climate Change*. New York: Bloomsbury.

Leonard, Christopher. *Kochland: The Secret History of Koch Industries and Corporate Power in America*. New York: Simon & Schuster.

Mann, Michael. *The New Climate War: The Fight to Take Back Our Planet*. New York: Public Affairs.

Rich, Nathaniel. *Losing Earth: A Recent History*. New York: MCD.

Speth, James Gustave. *They Knew: The US Federal Government's Fifty-Year Role in Causing the Climate Crisis*. Cambridge, MA: MIT Press.

What Lies in Store

Akkad, Omar El. *American War: A Novel*. New York: Knopf.

Stanley Robinson, Kim. *Forty Signs of Rain*. New York: Bantam Dell.

Wallace-Wells, David. *The Uninhabitable Earth: Life After Warming*. New York: Tim Duggan Books.

Index

INDEX

INDEX